Virus vs Mankind

(The Coronavirus Pandemic 2019)

Dale Mark Kudsy

Grosvenor House
Publishing Limited

This book is published by
Grosvenor House Publishing Ltd
Link House
140 The Broadway, Tolworth, Surrey, KT6 7HT.
www.grosvenorhousepublishing.co.uk

A CIP record for this book
is available from the British Library

ISBN 978-1-83975-779-2

Twitter and Instagram handle: **dkudsy**

ACKNOWLEDGEMENTS

A special thanks to God Almighty Jehovah who has looked after me and my family during this uncertain and dangerous time. Thanks to my lovely parents Emile and Fawziyah Lyras, my beautiful and supportive wife, Janet, and our two beautiful girls, Jasmine and Fawziyah. During the lockdowns in the UK, these were the only people I saw on a daily basis. These were the only people I had a face-to-face conversation with. They were my bubble. They helped to make the lockdowns easier and less difficult compared to others out there. My wife and my mum were in the kitchen trying to make us delicious meals as often as possible to keep us sustained while me and my dad drove around once every few weeks trying to restock our food supplies and all other household essentials. After eating, with nothing else to do and nowhere to go, we would watch the News to keep tabs on the Coronavirus pandemic in the UK and around the world. After watching the News, we would then watch movies on Netflix or on Amazon Prime to take our minds off the pandemic. In the morning after breakfast, I studied the kids to help them stay on track with their curriculum.

DEDICATION

This book is dedicated to all those who fell victim to COVID-19, both the deceased and the survivors and their families who have been through hard times, and to all those who have been fortunate enough not to catch the disease but have lived through all the tough life changing restrictions and sadness. It is also dedicated to our brave men and women of the NHS, and to all the frontline medical workers and scientists around the world who are risking their lives and their family lives to save others and to find a cure for the disease. It is also dedicated to all the other key workers such as delivery drivers, post men and women, supermarket staff, teachers, police officers, soldiers, shop keepers, carers, funeral undertakers, bus and train drivers, bin collectors, volunteers in all sectors, COVID-19 aid groups, and much more. All those who had no choice but to continue working to keep vital services running so that all those who are in lock down or are vulnerable, will be able to survive the pandemic and sustain themselves in their homes. We say, THANK YOU! Thank you for putting your life on the line for humanity and your country.

DISCLAIMER

Majority of information found in this book came from my hard work and from the thorough research I have carried out on the pandemic. However, some of the information, like statistics on the pandemic and News articles that are vital for this book to be able to paint a clear picture on the pandemic and all that has transpired, have been properly referenced, and are the hard work of the authors.

CONTENTS

INTRODUCTION

In the realm of infectious diseases, a pandemic is the worst-case scenario. When an epidemic spreads beyond a country's borders, that is when the disease officially becomes a pandemic.

It is March 2020, and fear looms in on the world. One with the potential to wipe out half of the world's 7.8 billion population. A fear so deadly that it will grind the world's economy to a halt and put every major financial capital and country on a lockdown, claiming an unprecedented number of lives. A fear which the 21st century generation is totally prone to, but has never envisaged, nor prepared for. A fear known as COVID-19!

COVID-19 is believed to have started in December 2019 in Wuhan, China, but it took several months for the rest of the world to notice it. Wuhan is an epicentre and a sprawling city of central China's Hubei province. It is a commercial centre divided by the Yangtze and Han River. The city has 11.8 million people with lakes and parks, including the expansive, picturesque East Lake. Nearby, the Hubei Provincial Museum displays relics from the Warring States period, including the Marquis Yi of Zeng's coffin.

Some people believe that COVID-19 could be the beginning of the long awaited Biblical prophesied Apocalypse, which is said to usher in the end of the world as we know it. Others, however, believe that this outbreak is the catastrophic result of an error that occurred in a nuclear bunker somewhere in the Wuhan province of China belonging to either America or the United Kingdom. Some also believe that it is a deliberate act of genocide from either the West or its enemies. Enemies such as Iran, ISIS (Islamic State of Iraq and Syria), and much more. This suspicion on Iran and others being responsible for the deadly COVID-19 virus is down to the fact that ever since President Donald Trump took office in the United States on January 2017, he has been targeting Iran, North Korea and other countries in so many ways that it almost resulted in World War 3. Could this outbreak be the revenge from these countries gone wrong? No one knows, it is merely speculation. Other people believe that the COVID-19 pandemic was caused by the installation of 5G mast mobile communication towers around the world. These people, including celebrities like Wiz Khalifa, Keri Hilson, Woody Harrelson, Amanda Holden, Amir Khan and many more, believe that 5G radiation weakens people's immune systems, making them more susceptible to catching COVID-19.

Think about it, if there was no truth to what these people are saying, then why are the governments, social media companies and various groups trying to shut these people down by deleting their posts on social media and classing their posts as conspiracy theories? What are the governments so afraid of that they do not want their people to know about? What are they trying to hide? Are they not the ones who said everyone should have the right to their opinion and that people should have freedom of speech? So, why don't they want these people, who are bringing all of these points to our attention, to have freedom of speech?

So just what is COVID-19 and why is it so deadly? COVID-19 is the abbreviated form of the 2019 strain of the Coronavirus Disease. Coronavirus has broken out before in the form of SARS, but nothing of this nature. This strain, dubbed as **COVID-19**, is totally unknown to the world and medical experts. Coronaviruses are a group of viruses that can cause illnesses ranging from the common cold to more severe diseases like Middle East Respiratory Syndrome (MERS-COV) and Severe Acute Respiratory Syndrome (SARS-COV). The virus is believed to be a zoonotic disease, which means it is transmitted between animals and people. There are many Coronaviruses that are circulating in animals, but this strain is the first to have invaded the human immune system in this way. Some symptoms of COVID-19 can be identified as respiratory symptoms, high fever, constant dry cough, shortness of breath and breathing difficulties. In the worst-case scenario, pneumonia, Severe Acute Respiratory Syndrome and kidney failure are the signs of the virus at its deadliest stage and in most cases, death is the sombre answer to its end. The new Coronavirus spreads primarily through droplets generated when an infected person coughs or sneezes, and someone else inhales it or the droplets enter through the eyes, mouth or nose of someone else.

From the Black Deaths of 1346, which also started somewhere in China, and claimed 75 to 200 million lives worldwide, to the Spanish flu, also known as the 1918 Flu Pandemic, which was an unusually deadly influenza lasting from January 1918 to December 1920. It infected 500 million people, about a quarter of the world's population at the time. The world has seen its fair share of viral pandemics. Then, in spring 2009, the H1N1 Swine Flu broke out infecting as many as 1.4 billion people around the world and killing 575,400 people by the time it came to an end in spring 2010. This was then followed by the 2014 Ebola outbreak in West Africa killing 11,323 people in general. Now, with the new Coronavirus in more than 170 countries worldwide and counting, and with 7.8 billion people living in the world, just how dangerous is life now?

On 28th February 2020 people in the UK were already taking notice of the outbreak. It would have been difficult to ignore entirely the headlines about what was happening in China, South Korea, Iran, and Italy. The first confirmed cases among British travellers returning to the UK had come as early as January 2020, but it still seemed possible to regard this as something happening, for the most part, a long distance away. Not every front-page newspaper that Friday morning led with COVID-19. The Daily Mail splashed on the saga of Prince Harry and his wife Meghan Markle (Duke and Duchess of Sussex). The Daily Express was leading with Brexit talks as the UK had just signed Article 50 to leave the EU, but most of the other newspapers did cover the deadly virus. In the final week of the month, 442,675 phone calls were made to the non-emergency NHS helpline 111. People were not yet panicking, but a generalised sense of low-level anxiety was everywhere.

CHAPTER 1

BRIEF HISTORY OF PANDEMICS AND HOW THEY CHANGED THE WORLD

Cholera, Bubonic Plague, Human Immunodeficiency Virus (HIV/AIDS), Smallpox, and Influenza are some of the most brutal killers in human history. Outbreaks of novel diseases spreading across international borders when there is no immunity in society to combat them are properly defined as pandemics, with Smallpox, which throughout history has killed between 300–500 million people in its 12,000-year existence, are revered as one of the worst pandemics ever.

With a shift to the Neolithic era (New stone age), which is also known as the agricultural or (Agrarian) revolution, 12,000 years ago in Britain, and around the world in the mid-17th and 19th centuries, which created farming communities that made epidemics more possible; outbreaks of Malaria, Tuberculosis, Leprosy, Influenza, Smallpox, and others first appeared during this period. As populations grow and cities around the world are developed, along with the modernization of societies and the fact that human behaviours have altered, pandemics have erupted ferociously. It was due to the change of human behaviour during the agricultural period which enabled people to start developing new ways of getting food or working in order to grow and harvest crops which were in demand then, that created new national and international public health crises like the current COVID-19 pandemic.

The impact of pandemics today and in the past is always unequal. For example, the impact of Smallpox on indigenous communities; some groups are hit much harder than others. The impact is also spatially uneven, generally exacerbated by existing racial, class and gender inequalities. Some pandemics, such as Ebola or MERS, wreaked havoc in specific areas or in a limited number of countries but were contained before they could spread to other parts of the world. Early containment of pandemics in the 20th and 21st centuries partly emerged from previous experiences of uncontrolled diffusion. The diffusion of the 1968 Hong Kong influenza pandemic shows how the swift global spread of influenza between 1968 and 1970 led to a reformulation of global health policies that emphasized the need for enhanced preparedness for future outbreaks.

The Black Death (also known as the Pestilence, the Great Mortality, or the Plague) was a bubonic plague pandemic occurring in Afro-Eurasia from 1346 to 1353. It is the most fatal pandemic recorded in human history, causing the death of 75 to 200 million people in Eurasia and North Africa. According to History.com, the plague arrived in Europe in 1347 when 12 ships from the Black Sea docked at the Sicilian port of Messina. People gathered on the docks were met with a horrifying surprise: most of the sailors aboard the ships were dead, and those who were still alive were gravely ill and covered in black boils that oozed blood and pus. Sicilian authorities hastily ordered the fleet of "death ships" out of the harbour, but it was too late. Over the next five years, the Black Death would kill more than 20 million people in Europe, almost one-third of the continent's population at the time.

By the time the plague wound down in the latter part of the 19th century, the world had utterly changed. The payments of farmers and craftsmen had increased and even tripled for some, and nobles were knocked down a notch in their social status. The church's reputation in society was ruined, and Western Europe's feudal system was on its way out. An inflection point that opened the way to the Reformation and the greater worker gains of the Industrial Revolution and beyond. The world had become more interconnected than it had ever been. Despite the two world wars that came many years later, the world recovered from the pandemic, on steroids, within the next three decades: a period of massive globalization in which manufacturing parts seemed to come from everywhere and undergo assembly anywhere.

Thought to have killed perhaps half the population of Europe, the Plague of Justinian was an outbreak of the Bubonic Plague that afflicted the Byzantine Empire and Mediterranean port cities, killing up to 25 million people in its year long reign of terror. Generally regarded as the first recorded incident of the Bubonic Plague, the Plague of Justinian left its mark on the world, killing up to a quarter of the population of the Eastern Mediterranean and devastating the city of Constantinople, where, at its height, it was killing an estimated 5,000 people per day and eventually resulting in the deaths of 40 percent of the city's population.

The Spanish Flu has the reputation of being one of the deadliest disasters in the history of mankind after the Black Death. Though it is called Spanish Flu, the flu did not originate in Spain. It was called Spanish Flu because the Spanish press was the first to report it as Spain was not involved in World War I, their press was free to report the news. Emerging at the end of World War I; the 1918 flu pandemic was a great contributor to instability not only in Europe but around the world. The month of peak mortality in the pandemic was in November 1918, the same month that the war ended. The US Navy and Army estimated that 40 percent and 36 percent of their servicemen had been affected. The descriptions of this pandemic disease are gripping, and the pattern of who was struck was different than in previous outbreaks.

The flu lasted for two years, and between the first recorded cases in March 1918 and the last cases in March 1920, an estimated 50 million people were dead, though some experts suggest that the death count might have been twice that amount. The Spanish Flu killed more people than the First World War, and possibly more people than even the Second World War, or even more people than both wars combined if these experts are correct. The flu spread in three waves: the first is in the spring of 1918, the second and most deadly was from September 1918 to January 1919, and the third from February 1919 through to 1920. The first two waves were intensified by the final years of World War I. The flu was particularly deadly to young adults without any underlying health conditions, which increased its economic impact relative to a disease that mostly affects the very young and the very old. The flu had been made deadly through a mutated virus which was distributed by wartime troops. According to nber.org, the pandemic reduced real per capita GDP by 6 percent and private consumption by 8 percent, declines comparable to those seen in the Great Recession of 2008–2009. In the United States, the flu's toll was much lower: a 1.5 percent decline in GDP and a 2.1 percent drop in consumption.

The decline in economic activity combined with elevated inflation resulted in large declines in real returns on stocks and short-term government bonds. The pandemic literally destroyed the economy of every country it struck. According

CHAPTER 2

LIFE BEFORE COVID-19

With 7.8 billion people living in the world in the 21st century, and technology developments spiralling out of control, with companies like Apple, Samsung, Microsoft, Boeing, Airbus, Tesla, Ford, Bentley, Space X and others, all competing to develop the next generation of mobile phones, TVs, computer, cars, aeroplanes, space shuttles, 5G radio frequency, smart watches and so on, humanity is repeating history once again by creating a breathing ground for vicious viruses like the Coronavirus, Cancer, HIV, Ebola and so on, just like they did during the New Stone Age. With the creation of motor vehicles' exhaust pipes and aeroplanes exhaust pipes, fossil fuel power plants, as well as some agricultural tasks or methods that farmers use, air pollution was born, which in essence is expelling dangerous gases into the air and into our lungs whenever we inhale them unknowingly. Also, nuclear weapons being developed by the powerful nations in the world like Britain, the United States of America, Iran, China, Russia and so on, for the eradication of potential foes, but people forget that the entire human race breathes the same air as it is circulated throughout the world, which means that humanity could be wiped out if a nuclear weapon is launched intentionally or accidentally. Even those launching it will fall prey to the chemical weapon once they run out of man-made oxygen in their bunker and start breathing in the natural air which has been contaminated with a poisonous gas. All of these could have created the COVID-19 pandemic which has now ravaged mankind and the world as we know it.

In the space of a month, the United Kingdom has changed beyond recognition. Most of the public has not got a clue about the transformation that was about to take place. One night when the UK was already falling apart due to Brexit, a chalkboard leaned on a lamp post outside a pub. Its message was an eerie one. “Unfortunately, the pub is closed. A customer who visited us has tested positive for the Coronavirus and as a precautionary measure, we are closing for a full deep clean.” This customer was allegedly known to have been the UK’s first case of COVID-19 that was caught inside the country. On the same day, 28th February 2020, grimmer news arises. A British man who had been infected on the Diamond Princess cruise ship has become the first UK citizen to have died in Japan from the virus. That afternoon in the UK, everything was happening as normal. Children were still in their classrooms undertaking their daily curriculum and adults were at work typing a few emails here and there and talking on the phone with clients. People shook hands and hugged and kissed each other. At night-time, they head for the pubs, clubs, and restaurants. Some go on a date night while others visited families and friends. Others go to watch a new released movie at the cinema. They queue up close to each other to buy their tickets, and when it is their turn, they get up close to the cashier behind the counter to order their ticket and a drink and some popcorn. Some will pay by cash and others will pay by card. The choice is theirs. As they make their way to the screening rooms, they walk up close to each other and when they sit down to enjoy the movie, people sit very close to one another. If someone coughs next to you, you are not bothered. It is just a normal cough.

As the week goes by, football fans across the country gather outside stadiums to buy their tickets or to go through security checks. Once the checks are over, some people talk to others or queue up again to buy themselves a cold pint of beer to get the match spirit going. Some decide to go for a quick visit to the toilets to ease themselves before the game begins. On their way into the toilets, they open several doors, and when they finish using the loo, they squeeze a drop of sweet-smelling soap onto their palms by pressing the squeezer on the hand wash followed by opening the old-fashioned taps to wash their hands by twisting its head with their hands, of which many other fans have touched. Coming out of the toilets, some people go straight to the stand which is marked up on their tickets while others buy themselves a delicious snack or sandwich from the various food vendors within the stadium. While at the stand, when the game is on the way, people are cramped tightly together, some sweating whilst others make friends and shake hands to celebrate their team's goals or advantages.

People go shopping to buy their favourite clothes or electronic gadgets from popular stores full of customers. They mingle with each other as they stroll down the shopping isles. They try clothes on and off and pick up electronic goods like laptops, game consoles, mobile phones to examine them before buying. Shelves are full of all sorts of things that people may want. Toilet rolls, hand sanitizers, Dettol, handwash, pasta and rice were plentiful. No one took these items as important as they do now during COVID-19. Some people have never even bought a hand sanitizer or a mask in their life before. People are free to buy as much toilet roll, hand sanitizer, rice or pasta as they want without any restrictions. When they have made up their mind on what they want, they confidently walk to the tills with their goods in the shopping basket or in the trolleys and they then queue up behind each other, almost jamming into the person in front or behind. When they get to the front of the queue to check-out, the cashier handles their goods with bare hands while talking to the customer with no protective screen in between them. Some cashier even helps the customers with packing the goods into their carrier bags to help speed up the check-out. The prices for goods are reasonably low for what you want to buy. Retailers and supermarkets even offer discounts to encourage people to buy more goods in order to get rid of older stock, allowing room for new stock on their shelves. Deliveries are regular. Supplies get delivered when needed by the stores. Deliveries come from warehouses across the country, and cash is accepted throughout the country.

Whether it is on Friday, Saturday or Sunday, religious worshippers from across the country gathered outside their place of worship for their weekly or daily service and devotion. Muslims at their mosque, Christians at their church, Jews at their synagogues and so on. Throughout the week, Muslims congregate in front of their local mosques to pray. Some Muslims visit the mosque five times a day while others only visit once a day or once a week. Upon entering, Muslims take off their shoes and leave them on a shoe rack and then they hug and shake hands with one another. They then proceed to wash their hands and feet before entering the main prayer hall. In the prayer hall, they sit close to each other as they listen to their religious leader, the Imam, as he delivers the service. The prayer halls of the mosque are fully carpeted so that worshippers can just enter and sit wherever they wish.

Christians on the other hand, whether it be the Catholics, Pentecostals, Born-Again, Anglicans and so on, all have different ways of worshipping. For instance, on Saturdays, Sundays, and other days of the week, Catholics congregate outside the Roman Catholic Church for their weekly service. Hundreds of people from all sorts of backgrounds meet and greet each other outside the church with hugs and handshakes. Congregation members dip their fingers into the large water font to make the sign-of-the-cross before entering the church. When the service begins and the church is congregated, worshippers listen attentively to the Priest as he delivers the mass. Just before the Offertory is collected, a peace-be-with-you is initiated which is done by shaking each other's hands in the church. This is then followed by the giving of the Holy Communion. Worshipper's line up row by row behind each other as they proceed to the front of the church to receive the Holy Communion and wine. Communion is given by the Priest and several volunteers from the congregation into the hands, or directly into the mouth of worshippers who may be carrying a child. Worshippers then queue up to receive the blood of Christ by drinking red wine from the same cup as others before them.

Freedom of life is visible from around the country in this 21st century era. People visit their local parks with their kids, or alone, if the weather is nice and stay as long as they want. Their kids play freely with other children in the park. They share the same play area and swings and sea-saws. With the weather permitting, some people visit the beach in their shorts. They sit on the sand surrounded by many other people as they relax and enjoy the breeze while eating an ice-cream or a sausage roll. Some jump into the sea for a swim and their kids play in the mini pool which is filled with people. Exercise enthusiasts go to the gym regularly to exercise. They share the same exercise cardio machines, treadmills, exercise bikes and muscle building weights, all of which are draped in sweat from various users. People move freely as they wish. They use these cardio machines and weights without needing to wipe them down thoroughly before use.

Down in London, everything is going smoothly. The London Underground Tube and buses are running as planned and are jam packed during the rush hour periods in the morning and evening. People are travelling from one end of the city to another. Perhaps the only fear in the back of some people's mind is the fear of knife crime or a terrorist attack, which has recently spiralled out of control in the capital and across the country. People were scared to leave their houses, not knowing if they would be the next stab or terror victim, a mere statistic for the news, as lots of people have been stabbed to death in the capital city, with terrorist attacks also being quite high. Apart from that, people were happy and travelled freely. They travelled back and forth on the underground and buses daily. In the early morning hours, men and women in suits, carrying a briefcase or a rucksack with a laptop in it, traverse the city in a rush to catch their train or bus, not wanting to be late for work or late for that important business meeting. Construction workers, who are casually dressed with paint all over their clothes and a hard hat in one hand and a tool bag in the other, also make their way to work using the public transport system. The world-renowned Oxford Street is crowded with shoppers as always, workers and tourists walking tightly close to each other, looking to buy top brands of fashion and electronics. People are also flocking to popular tourist sites like Buckingham Palace, Big Ben, the Shard, Tower of London, London Eye and many more, to enjoy themselves. All of this was a normal daily lifestyle for people until COVID-19 interrupted it all and sent the world into fear and isolation.

CHAPTER 3

LIVING IN FEAR

By 1st March 2020, the virus had reached the four corners of the United Kingdom. Cases had been detected in England, Northern Ireland, Scotland, and Wales. Two days later, with the total number of confirmed cases at 51, Prime Minister Boris Johnson stood behind a lectern and launched the government's Coronavirus Action Plan. The outbreak was declared a '**level four incident**'. The following day, the government's SAGE committee of scientific experts was shown a revised modelling of the likely death toll. The figures, according to the Sunday Times, were 'Shattering'. If nothing was done, there would be 510,000 deaths. Under the existing 'mitigation' strategy, shielding the most vulnerable people while letting everyone else go about their business mostly as normal, there would be a quarter of a million deaths in the country if a more draconian type of action was not taken. In a press conference, the prime minister told the public that anyone with a continuous cough or a fever must self-isolate. His instruction came with a warning that, "Many more families are going to lose loved ones before their time". The bluntness was shocking. Some asked, why wasn't more being done.

On Friday 13th March 2020, the London Marathon, the Premier League and English Football League and May's local elections were all postponed. Scotland had its first Coronavirus-related death. Saturday 14th and Sunday 15th March was the last relatively normal weekend. You could not watch league football, but you could go to the pub. Hand sanitiser could not be found on any supermarket shelves, but you could tell your friends about your plans to practise 'social distancing' if you met them on the street. Around the country, people looked at Italy, France and Spain, which had already gone into national lockdowns, and wondered if the UK was next. Volunteers began forming mutual aid groups to deliver food and medicine to vulnerable people who were self-isolating or shielding.

On Monday 16th March 2020, Boris Johnson advised against non-essential travel, urged people to avoid pubs and clubs and work from home. Across the country, kitchen tables were cleared to make way for laptops. Thanks to Skype and the virtual meeting app, Zoom, workers started getting a glimpse of their colleagues' interior decor. Those who could not work from home wondered how on earth they were supposed to earn money and stay safe from the virus. On 17th March, the government began holding daily press conferences, events that would soon become regular viewing for nervous families. Just six days after presenting his budget, the Chancellor, Rishi Sunak, announced £300 billion in loan guarantees, a huge expansion of state intervention in the economy by a Conservative government.

There were still calls for more to be done to stop Britons infecting each other. The following day, school pupils, apart from those whose parents were not designated key workers, were told they would not go back to their classes until further notice. GCSE, SAT, CATS, A-LEVELS exams, proms, farewells to classmates and teachers, would now not happen for a while, or probably not happen at all. All through the following week, people would look forward to their one state-sanctioned form of outdoor exercise a day. They would stand in front of their laptops, following the instructions set out by the fitness coach, Joe Wicks, and others on YouTube.

Just imagine, the sun is shining, spring is around the corner, and you have money in your bank account. You have been given one of the longest leave of absence from work ever and your kids are officially off school, even before the Easter holiday begins, and no one knows when they will ever resume again, and yet, you are locked inside your house afraid to trespass outside. You have even become afraid of venturing into your own porch or garden, all because of an invisible virus that is lurking around the corner, or rather in the air.

Your normal way of living is now altered by fear! Fear of becoming sick, and even more worrying, fear of dying too soon. People begin to feel that their government has given them a death sentence, because their entire country and world in general is in lockdown. All shops are closed, apart from essential shops that sell food and medications. Despite being allowed to open, supermarket shelves are empty, even though the British government keeps saying to its people not to worry and that they have enough food supplies for the whole country. Shelves are empty because people are panic buying as no one knows when the lockdown will end and when they will be allowed to go outside again. It could be weeks or months, or even a year, before the curfew is lifted, and people can return to their normal way of life.

The British people grow anxious by the day. Some, by the hour. The fear of death becomes more and more imminent and real. If you don't die from the virus, you are most likely to die from hunger or thirst as food is being rationed, and for some, they could die from a mental break down due to the fact that they are being locked in for weeks and months, not able to go outside of their own house or see their loved ones. Not able to do the things that they love such as going to the gym, going for a morning run, visiting family and friends, going to watch your favourite movies at the cinema, going for dinner at your favourite restaurant or hanging out at night with your friends at the pub or club to have a drink and some fun. Life has become so restricted that human rights and freedom is now deemed as a joke.

With the UK Prime Minister and Health Secretary, Matt Hancock, also showing symptoms of COVID-19 and going into isolation, the whole country is in sheer panic. It has been more than a week since the Prime Minister stood at the podium in Downing Street to deliver the daily Coronavirus briefing. Since announcing he had tested positive for the virus, his public appearances have been limited to video clips filmed on his phone from inside his flat, above 11 Downing Street. In his latest self-shot video message, he said he was still showing mild symptoms of the virus and would therefore stay in isolation until they passed. In a direct plea to the public, he urged people to stick to social distancing guidelines. Matt Hancock had also tested positive for the virus and returned from self-isolation to host the daily Downing Street news conference. A few days later, Boris Johnson was taken to hospital and by the next day he was moved to intensive care as his Coronavirus symptoms had worsened, with overall charge of the government handed to Dominic Raab, the Foreign Secretary. It was said by a Downing Street spokesperson, that he was on oxygen and that the move was just for precautionary reasons. On 12th April, he was discharged from hospital and taken to Chequers to recuperate, until he was once again strong enough to resume his prime ministerial duties.

A few weeks before the pandemic began, Andrew Parker (21 years old), from London, would not have thought twice about mingling in a night club full of sweaty revellers, or going to the Notting Hill Carnival in August. Parker, just like most young men and women, was an avid clubber. He enjoys the outdoors and the night life of England. He enjoys going out to different clubs and pubs with his friends, full of people from all different walks of life. Parker and his friends would enjoy each other's company, the ice-cold alcoholic drinks, and the disco music that the DJ plays from his or her powerful sound system. They took COVID-19 for a joke when it started, saying that they have even drank a beer name Corona Extra before and no one died. They used to say that COVID-19 is just like ordinary cold, and that the media is just trying to frighten the public.

According to Parker, the Flu has killed more people than Coronavirus can ever achieve. So why panic? It was when the death counts started increasing in the UK and the number of people who had tested positive for COVID-19 rocketed, and the government started restricting people's movements and ordering people to only go out to buy essential things like medicines and food, that Parker and his friends began to take Coronavirus seriously.

Parker is a regular at his local gym and a fitness fanatic. With gyms closed all over the country, Parker is unable to do his daily visit to the gym to exercise. He feels bored being locked in his house and must find other ways to keep himself fit. He is now doing sit-ups, press-ups and other exercises in his tiny living room, using online training videos from YouTube to guide him. Before COVID-19, whenever Parker went to the gym, he would go with his friends Eddie and James. Now, however, Parker is locked inside his house, afraid to open the door to his once best friends in case they have the lethal virus. Even his girlfriend, Ava, who he used to cuddle up close to and kiss, is unwelcome in his apartment. Their relationship has now become a long distant one. They now communicate on the phone and WhatsApp video calls regularly. His only companion is his PlayStation, where he plays FIFA 20, Red Dead Redemption 2 and many other games to drain his fear of the virus. Even when the post comes through the letter box, Parker has to put on a disposable glove to pick up the letters and after reading the contents, he would bin the glove and wash his hands several times.

Andrew Parker is extremely paranoid about COVID-19 as he and millions of other people have never seen anything like it before, and the daily increase in deaths terrifies them, especially when the deceased is a young person with no underlying health conditions. This made Parker realise that this disease is not about age or how fit you are like the government says. Nor is the disease about race or the colour of your skin, or how wealthy or powerful you are. It has no boundaries, and it strikes ferociously at anyone who dare to stand in its way.

There is only one thing that takes Parker out of his apartment and that is to restock his food supplies. Whenever he has to go to his local supermarket, which is about 15 minutes away from his apartment, his heart starts beating extremely fast for a journey which he has made many times before. Parker ensures that he covers every exposed part of his body, such as his face and neck, by wearing a strong, tightly sealed mask, along with a Rayban to protect his eyes. He would wear his gloves, along with a scarf to cover his neck. It is as if he is practicing self-isolation while being outside.

When he goes outside, he would do his very best to not touch any surface or handles even though he is wearing gloves. He would keep himself far away from people as much as possible, and if he has to use a trolley, he would wipe the trolley handle properly with disinfectant wipes which he carries around with him.

Some shops and supermarkets are closed even though they are essential shops, maybe because their owner has fallen victim to the virus. People now have to queue up outside the shops until one or two other customers come out before others can enter. This is now the new norm, and very soon the whole world will get used to it. All the shops have a staff member at the entrance to supervise this social distancing, which most people find unusual and difficult to get used to as people are used to walking into any shop of their choice, full of other shoppers, without queueing up outside to wait for the shoppers inside to leave. Nowadays, when Andrew Parker comes across someone he knows, he will keep himself far away from them as he talks with them. When he goes home, he will wash his hands and sanitize them properly before washing almost half of his shopping such as tin foods, and all the bottle drinks along with all the fruits and vegetables with soap and rinsing them properly.

The COVID-19 outbreak has already brought a range of challenges for the frontline medical staff. Shortages of equipment, doctors and nurses falling ill and thieves breaking in to steal valuable medical supplies. In Cambridgeshire, 55-year-old Amy is petrified because she works on the frontline of the NHS as an A&E Doctor at Addenbrookes Hospital, dealing with people infected with Coronavirus. She has no choice but to go to work as she has not demonstrated any symptoms of the disease and the government has instructed every key worker, such as medical staff, police officers, carers and delivery drivers to go to work unless they are infected or come across someone that tests positive for

COVID-19, then they can go into isolation. She is scared because a handful of her colleagues have caught the virus despite following all precautionary measures, and she could be next. She has even heard that one of her colleagues has died from the virus, and yet she must control her emotions and hide her fears and do her best to save the next COVID-19 patient's life before they become another death statistic for the News.

Amy and her colleagues feel vulnerable and exposed to the virus as their employer, the National Health Service (NHS) and the government has not provided them with appropriate PPE (Personal Protective Equipment). They feel vulnerable because the apron they wear is very thin and the slightest amount of infected saliva from a patient's cough or sneeze can get to their skin. The facial mask is not sealed tight enough to cover their mouth or nose, which leaves them exposed to any viral particles in the air or when talking to a patient. Their eyes are fully exposed as they do not have any eye protection to prevent the virus from going into their eyes.

Patients who she used to touch freely in order to examine them thoroughly can no longer happen. Now she has to wear an acrylic disposable glove to do her job. Her colleagues, who she used to work in close proximity with, must keep 2 metres apart from one another. Everyone is afraid of each other. The slightest cough that you cough, means that every eye will glance at you suspiciously and people will keep as far away as possible from you. Someone who was walking towards you will immediately alter their course so that they can keep far away from you.

How can you treat someone properly if you cannot get close to them? In the hospital wards, social distancing is impossible. Patients in high-risk groups are vulnerable during the incubation period as people may not show symptoms. In the wards, staff work closely with health authority guidelines by making sure that they wash their hands regularly, and by wearing a surgical mask, plastic apron and gloves for suspected and confirmed COVID-19 patients. Many staff who had contact with COVID-19 patients when providing treatment or care for them have no choice but to continue to go to work unless they display symptoms. There are two folds to this. Patients will infect staff or staff, who might be infected and not know it because they are not being tested, can also infect the very patients that they are supposed to treat, and this will mean that many more staff will be off sick or even dead, which will leave the hospitals with no health professionals to look after the public.

Amy is also worried because she lives with her 60-year-old husband who has severe Asthma, and he is deemed as vulnerable. She is worried that she might contract the disease from work and bring it home to her husband. She fears that her husband might not survive the disease if he gets it.

In Bradford Royal Infirmary Hospital, all 18 entrances across the 26 acres site are sealed off and everyone is now ushered in through one main entrance door and leaves through another. Staff must now show passes on demand and patients are no longer allowed a visit from their loved ones, nor are they allowed to bring someone along for appointments. Staff on the front line were even more frightened because they were not tested for the virus unless they were about to faint during their shift, which makes it extremely scary for them, their colleagues, and the immediate family members that they live with. Some staff, for a job that they love so much, dread it when their shift is at hand. Some even regret choosing a medical profession for their career.

A doctor in the hospital has started showing symptoms of COVID-19, and knowing how important he was to the department, they decided to test him. Later in the week, the results came back positive, and panic took over his department. More and more doctors and nurses in his department or within the hospital began to fall ill and call in sick so they could self-isolate. During the isolation, more and more fear set in. Not knowing if you would survive it or not. Not knowing how to treat yourself medically inside your confined space, even if you are a doctor, as nothing seems to work. The panic erupts inside you because you do not know whether if your rent or mortgage will be paid at the end of the month as your employer will not be paying you your full monthly salary.

With Coronavirus scattered all over the country, the fear is intensified each day as the UK government passes tougher sanctions on the population to stem the flow of COVID-19. People are extra cautious with the things they do. Nobody wants to go near hospitals in case they catch the virus. People will now only visit hospitals if it is a life-or-death situation. They would rather manage the pain themselves at home rather than going to the hospital.

Like in the case of 28-year-old, Joan, who is 34 weeks pregnant with her second child, she became scared when she visited the hospital for a routine scan with her boyfriend and was told that because of COVID-19, no one is allowed to accompany her anymore for her appointments, not even her partner. She was scared because she struggles to do things nowadays by herself due to being heavily pregnant and relies on her partner to help her, as even walking the shortest distance nowadays is a difficult task and without her boyfriend accompanying her, how would she cope? She also worries that when her time comes for delivery in few weeks' time, she will be alone in the delivery suite at the time when she needs all the support from her partner. Joan is also becoming paranoid whenever she has to visit the hospital, thinking that she could catch the deadly virus and infect her unborn child who could then turn out to be born disabled. Her paranoia is so much that she even worries that when her time comes to deliver her child and her partner is nowhere to be seen because he is not allowed to be there with her, maybe, just maybe, one of the nurses could inject her and her unborn child with the virus in order to reduce the population of England and save the government money on benefits.

Families can no longer bury their loved ones and they are left to mourn alone inside their homes as the whole country is on lockdown and no one wants to risk their life or pick up a fine to come and mourn with you. The stigma of the virus is real. Your once close friends or family members that used to visit you frequently are now scared of visiting you just in case you also have the virus. Weddings, christenings, baptisms, and many other ceremonies are cancelled.

CHAPTER 4

HOW COVID-19 ATTACKS THE BODY?

When an infected COVID-19 person expels virus-laden droplets either through sneezing or coughing and someone else inhales them, the novel Coronavirus, called SARS-CoV2, enters the nose and throat. It finds a welcome home in the lining of the nose. Scientists studying the virus found that cells inside the nostril are rich in a cell-surface receptor called angiotensin-converting enzyme 2 (ACE2). Throughout the body, the presence of ACE2, which normally helps regulate blood pressure, makes tissues vulnerable to infection, because the virus requires that receptor to enter a cell. Once inside, the virus hijacks the cell's machinery, making myriad copies of itself and invading new cells.

As the virus multiplies, an infected person may shed copious amounts of it, especially during the first week or so. Symptoms may be absent at this point. Or the virus's new victim may develop a fever, dry cough, sore throat, loss of smell and taste, or head and body aches. If the immune system does not beat SARS-CoV-2 during this initial phase, the virus then marches down the windpipe to attack the lungs, where it can turn deadly. The thinner, distant branches of the lung's respiratory tree end in tiny air sacs called alveoli, each lined by a single layer of cells that are also rich in ACE2 receptors.

Normally, oxygen crosses the alveoli into the capillaries, tiny blood vessels that lie beside the air sacs; the oxygen is then carried to the rest of the body. But as the immune system wars with the invader, the battle itself disrupts this healthy oxygen transfer. Frontline white blood cells release inflammatory molecules called chemokines, which in turn summon more immune cells that target and kill virus-infected cells, leaving a stew of fluid and dead cells pus behind. This is the underlying pathology of pneumonia, with its corresponding symptoms: coughing; fever; and rapid, shallow respiration. Some COVID-19 patients recover, sometimes with no more support than oxygen breathed in through nasal prongs. Others deteriorate, often quite suddenly, developing a condition called Acute Respiratory Distress Syndrome (ARDS). Oxygen levels in their blood plummet and they struggle even harder to breathe. On X-rays and computerized tomography scans, their lungs are riddled with white opacities where black space air should be. Commonly, these patients end up on ventilators and the majority of them die. Autopsies show their alveoli became stuffed with fluid, white blood cells, mucus, and the detritus of destroyed lung cells.

It seems like this virus has no limits on where it can attack the human body. From the brain right down to the lungs, to the intestines further down the body, this virus has no boundary. In early March, a 71-year-old Michigan woman returned from a Nile River cruise with bloody diarrhoea, vomiting, and abdominal pain. Initially, doctors suspected she had a common stomach bug, such as Salmonella. But after she developed a cough, doctors took a nasal swab and discovered she was positive for the novel Coronavirus. A stool sample positive for viral RNA, as well as signs of colon injury seen in an endoscopy, pointed to a gastrointestinal (GI) infection with the Coronavirus, according to a paper posted online on the American Journal of Gastroenterology. Her case adds to a growing body of evidence suggesting the new Coronavirus, like its cousin SARS, can infect the lining of the lower digestive tract, where the crucial ACE2 receptors are abundant. Viral RNA has been found in as many as 53 percent of sampled patients' stool samples. The presence of the virus in the GI tract raises the unsettling possibility that it could be passed on through faeces.

The intestines are not the end of the disease's march through the body. For example, up to one-third of hospitalized patients develop conjunctivitis with pink, watery eyes, although it is not clear that the virus directly invades the eye. Other reports suggest liver damage: more than half of COVID-19 patients hospitalized in two Chinese centres had elevated levels of enzymes indicating injury to the liver or bile ducts.

COVID-19 affects people differently. Just look at the case of Sandra Johnson from New York. She is a very healthy 38-year-old with no health conditions whatsoever. Sandra loves her morning run before going to work as a police officer. The first three days after catching COVID-19 she had fever, body ache, back and neck pain, and a terrible headache. Then she felt better for a day. She was happy with her improvement. On day 5 she woke up and could not smell or taste anything. Coffee tasted like water. Food tasted like nothing. She could not smell perfume, shampoo, or anything else for that matter. She then developed a dry cough on day 6. A simple task like walking down the hallway of her house was a challenge. She finds it very hard to do anything. Despite being so sick, she forces herself to get up and cook and tend to her kids. Day 14 and she would like to say that she feels better but her breathing is off. Deep breaths still hurt like hell.

Like in the case of Sandra, Elaine Fisher from Cambridgeshire, England, who is a key worker in a care home, contracted COVID-19 after attending to one of their elderly residents who later died after testing positive for the deadly disease. Elaine came home that evening after a difficult twelve-hour shift and started displaying prominent symptoms of COVID-19. She lives alone in a two storey two-bedroom mortgaged house, with the toilet and bathroom downstairs and bedrooms upstairs. Elaine ran into trouble when she went to have her bath after a long tiring shift when she started feeling sick. She realised that she had a high fever along with a persistent cough and immediately became worried. A day later she became weaker and could hardly walk and had to call into work sick. Her employer told her to self-isolate for fourteen days. Days later, Elaine had to literally drag herself around her house to go to the toilet downstairs from her bedroom where she stayed most of the time. Her vision is blurry because her eyes are constantly watery, and her feet feel like jelly if she manages to lift herself from the floor. If she manages to make a drink or a snack, she finds it very painful to eat or drink it because her mouth has no taste, and swallowing is now a very difficult task due to the fact that her throat feels like it has multiple razor blades stuck to it.

In Brescia, Italy, a 53-year-old woman walked into the emergency room of her local hospital with all the classic symptoms of a heart attack, including tell-tale signs in her electrocardiogram and high levels of a blood marker, suggesting damaged cardiac muscles. Further tests showed cardiac swelling and scarring, and a left ventricle, normally the powerhouse chamber of the heart, so weak that it could only pump one-third of its normal amount of blood. But when doctors injected dye in the coronary arteries, looking for the blockage that signifies a heart attack, they found none. Another test revealed why: the woman had COVID-19.

Jane Mansfield, a healthy 45-year-old lady from Peterborough, England, began to feel more tired than normal and by the time she went to bed later that day, she was exhausted. That was a particularly tough weekend for her, especially as she is a single mum and lives with her 6-year-old son. A few days later, she started getting pains in her legs, which became excruciating. She thought it was a trapped nerve and took some painkillers, but to no avail. Later that day after a visit to her GP, the doctor phoned her and told her that she had COVID-19 and that the virus had gone directly into her muscles. She began to cough but it was not persistent like experts suggests. Jane's fear was that she might have infected her son and asked to get him tested. The son was picked up by Jane's brother and was taken to live with him far away from his mum when his test results came back negative.

Jane was bed-bound for over a week and decided to go to her local shop which is five minutes down the road to get some provisions to last her for a few weeks, and that was when it hit her. When she got back home from the shop that afternoon, she was freezing cold and shivering uncontrollably. At one point, she managed to make herself four hot water bottles which she wrapped around herself while she was covered under a thick blanket in order to keep warm on the sofa, but she just could not get warm enough. Then the fever got worse. She felt like her body was on fire, along with a splitting headache. Jane could not eat anything, instead she was vomiting and wringing wet in sweat, and then breathing started to become more difficult.

Being asthmatic, she became worried and thought she could fight the virus at home. Within a few more days, she was slipping in and out of consciousness and with difficulty, she was able to pick up her phone to call 111. The paramedics arrived and she remembers hearing the ambulance driver outside saying on the radio: "She's very poorly, we need to bring her in." He put an oxygen mask on her and carried her out to the vehicle. When they arrived at the hospital, they were placed in a queue of ambulances just waiting to off-load patients at A&E. She was lying there for about three hours until it was her turn. They put her in a wheelchair, and she remembers them saying they had no cubicles, and they were full to capacity. Jane sat there with her eyes closed listening to everything going on around her: people rushing around, phones ringing, general commotion all over.

The nurse said: "I have to swab you for COVID-19." He stuck the swab stick so far down the back of her throat that she was retching, and then just as she was recovering, he said: "Now I have to do it up your nostrils." That was followed by a raft of blood tests and a chest X-ray.

Jane felt pummelled. All she could think was, "What the hell's going on?" She felt like passing out. She remembers another nurse coming over and telling her: "Just to let you know, your X-ray results have come back, and you have got pneumonia in the lungs, and you'll have to be on oxygen 24/7."

At one point, she felt the most-almighty pain in her chest, like she was being compressed with slabs of concrete. The doctors told her it was the pneumonia attacking her lungs and they gave her a shot of Morphine. That was followed by terrible stabbing pains in her stomach, as bad as labour contractions, and she cried out: "I can't take this anymore! I can't carry on like this!" By the time the pain subsided, she was almost delirious. There were only four beds in her bay, and everyone in there had tested positive for COVID-19, and had an underlying health issue. Two other women in there were diabetic, and a third woman was brought in opposite her after a couple of days.

She does not remember much of the first few days, just nurses coming in and out all the time, and cleaners coming in to disinfect everything. Most of the noise came from Jane ringing the bell constantly and gasping for drinks of water. She was so weak that all she could say was, "Commode!" Despite being so weak and poorly, she was watching the nurses and they were all working a minimum of twelve-hour shifts. "You could just see that they were exhausted."

One night, Jane saw a man in what was meant to be an all-female ward. She rang the bell, and the nurse came and explained that he was the son of the woman in the bed opposite her and that she was an "End-of-life" patient. Jane became dreadfully sad for them but at the same time she was thinking, "So I've got somebody who is about six feet from me who is basically waiting to die and I'm going to hear it all unfold." They had the curtains pulled round each bed so that they can have a modicum of privacy.

That was when the hallucinations began. She began getting flashbacks of conversations she has had in her life so far and people she had met. At one point she thought, "Am I alive or dead? Do these flashbacks mean that I am transitioning to death? Is this what people mean when they talk about your life passing before you when you die?" And then I started saying repeatedly, "No, I don't think I am dead, because there is no white light and no angels and nobody calling me."

Suddenly it was the early hours of the morning, and Jane heard a male nurse outside the door saying, "She is gone!" The poor woman opposite her had died. Jane waited for them to come in and remove the woman's body, but nothing happened. That lady's body was there for what seemed like hours before they eventually came in. They were cleaning it and then wrapping it in plastic, like packaging. Then she heard them put the body in a body bag, zip it up and say, "On the count of three. One, two, three!" The noise of that woman's dead body coming into contact with the metal trolley, that is a sound that you do not forget for the rest of your life.

The cleaners came in and started cleaning where the woman had been and sprayed lemon scent to try to freshen up the smell. By daytime Jane was just looking at an empty bed. The day before, she had been looking at somebody and now the bed was empty. That thought really affected her.

She started watching the woman in the bed diagonal to her slip into a coma, and she watched her daughter come and say desperately: "Mum, it is me! Mum, it is me!" It was pitiful because the woman had already gone into a comma. It sounds awful but Jane was waiting for her for two nights to die, which was very distressing. The woman next to Jane was getting better and she commented that they were in a bay where 50 percent had died, and 50 percent had lived and that they were on the lucky side of the room.

Jane Mansfield had fought to stay alive. After being almost ready to give up at the start, she had told herself, "No, I have got to carry on, I'm not going yet. I'm only 45 and I'm not ready to die, not just for me but for my son and my family and friends." My brother, James, had texted me constantly with pictures of my son, along with love and supportive messages, and that gave me the will to fight the virus.

Unfortunately, the comatose woman died after two days and again Jane heard the same process. The plastic, the zipping, the trolley, and the cleaning. What saved her life perhaps was one male nurse who said to her, "If the doctors say you are medically fit to go home, go! Do not make the mistake of staying in hospital because you feel a bit weak.

Believe me, I have seen it in this ward. Every patient who has been told by the doctors that they can go home and have instead argued that they do not feel 100 percent and want to stay one more night in hospital. Every one of them has contracted a secondary illness, because this is a high-risk COVID-19 ward, and they have all died."

That same day, they tested her blood, oxygen, and saturation levels and she scraped by. The doctor said to her, "You have just made it. I'm happy to discharge you." Jane was so excited; she finally was going home.

It was freezing cold outside. She only had a hospital gown and flip flops on, but she could feel the air on her face, and she was elated. Jane does not know the name of the female ambulance driver, but she was an angel. She had started her shift that morning at 06:00, and she was picking her up at 00:20. She had done an 18-hour shift. This is what these people are doing. It is not just the nurses and doctors. It is the people who are driving the ambulances. It is the paramedic crews. It is the woman at the desk doing the admin. It is the man coming in cleaning up after a dead body is removed. It is the porter taking it down to the morgue. It is the security guard keeping us and the hospital staff safe from heartless people who might want to hurt us during this sad and difficult time. They are the heroes of COVID-19.

Jane is now bed-bound for the next few weeks and the doctors said it could take three to six months to get over the pneumonia. She is registered on the government's website gov.uk as living in isolation, and they supply her food stuff regularly. The delivery is left outside her front door for her to pick up, and with the little strength she has, preparing any food on her own is very difficult. But that's the way it is. No one can come to her help during such treacherous times.

CHAPTER 5

LIVING WITH ANOSMIA AND PAROSMIA

Many people with COVID-19 are noticing that they are losing their sense of smell. As they recover, this sense usually returns, but for others it is a permanent loss. Some people are noticing that things smell differently than they used to. Things that are supposed to smell nice such as food, soap and their loved ones, now smell repulsive. The numbers with this condition known as parosmia, are constantly growing, but scientists are not sure why it happens, or how to cure it. According to a study conducted by the BBC, Prof Barry Smith, UK lead for the Global Consortium for Chemosensory Research, says another striking discovery is what he calls 'the fair is foul and foul is fair aspect of parosmia'. Parosmia is basically the distortion of smell or a reduction in the intensity of smell. "For some people, nappies and bathroom smells have become pleasant and even enjoyable," he says. "It is as if human waste now smells like food and food now smells like human waste." These unpleasant smells are often described as being like chemicals, burning, faeces, rotting flesh, mould, and much more. For some people they appear in response to specific odours, and for others they can be triggered by virtually any smell.

Parosmia can range from mild to severe and can be an incredibly debilitating and depressing experience for sufferers. So, what causes parosmia? The prevailing hypothesis is that it results from damage to nerve fibres that carry signals from receptors in the nose to terminals (known as glomeruli) of the olfactory bulb in the brain. When these regrow, whether the damage has been caused by a car accident or by a viral or bacterial infection, which in this case, it has been caused by COVID, it is thought that the fibres may reattach to the wrong terminal. There is also another form of Parosmia which is Phantosmia. According to Fifthsense, Phantosmia is the term for olfactory hallucinations, or phantom smells that appear in the absence of any odour. These can manifest as 'normal' smells for example, being able to smell garlic when there is no garlic present, and they can also be unpleasant. Parosmia and phantosmia are both classed as 'dysosmia', or qualitative disturbances of the sense of smell.

Charlotte Barnes ends up in tears whenever she tries to cook for her family of four. "I go dizzy with the smells. A putrid smell fills the house as soon as the oven goes on and it's unbearable," she says. The 52-year-old from East London has been living with parosmia for eight months and it makes many everyday smells disgusting. Onions, coffee, meat, fruit, alcohol, toothpaste, cleaning products and perfume, all make her want to vomit. Tap water has the same effect (though not filtered water), which makes washing difficult. "I can't even kiss my partner anymore," she says.

Gemma caught coronavirus in June 2020 last year and, like many people, she lost her sense of smell as a result. It briefly returned in August, but by September Gemma was rejecting her favourite takeaways because they reeked of stale perfume and every time something went in the oven there was an overpowering smell of chemicals or burning. Since the summer she has been living on a diet of bread and cheese because it is all she can tolerate. "I have zero energy and ache all over," she says. It has also affected her emotionally.

Other people experience Anosmia which is the complete loss of smell. According to Becky, "Although the anosmia was not nice, I was still able to carry on with life as normal and continue to eat and drink," she says. "This is referred to as cross-wiring and it means the brain does not recognise the smell and is perhaps programmed to think of it as danger."

The theory is that in most cases the brain will, over time, correct the problem, but medical specialists are reluctant to say how long it will take. Apart from waiting for the brain to adapt, there is no cure, though AbScent believes "Smell training" may help. This consists of regularly smelling a selection of essential oils, one after the other, while thinking about the plant they were obtained from.

According to BBC News, two sisters, Kirstie, 20, and Laura, 18, from Keighley, have taken this approach, though it took a while to work out how to do it while also living in harmony with their parents. The sisters had to run around the house opening windows when their parents came home with fish and chips on one occasion, "Because the smell is just awful," says Laura. Their parents, on the other hand, have been getting tired of the hot spices the sisters cook with, in order to mask unpleasant tastes, and to provide a hint of flavour. The most pleasant tastes are fainter than they used to be. The loss of smell has become one of the predominant symptoms of positive COVID-19 cases. Why, exactly, is not known; like everything else with COVID-19, it is new. "We just have the current data to go on," Dr. Sindwani says.

He also says that there are a range of data points available, but about 85 percent of COVID-19 patients experience some sort of subjective disturbance in their sense of smell. "It is estimated about 25% of COVID-19 patients lose their sense of smell for more than 60 days even," he adds. For long-term smell loss, that number is much smaller. "One study used objective smell testing and found that only 15% of COVID-19 patients experience a loss of smell for more than 60 days and less than 5% experienced it for longer than six months. That is really comforting news," Dr. Sindwani says in the Cleveland Clinic post.

CHAPTER 6

UK GOVERNMENT'S UNPRECEDENTED STIMULUS PACKAGE RESPONSE

As the death toll around the world increases to 68,413 and counting and the UK at 4,932 deaths and counting, big household names and brands like T.K Maxx, Debenhams, Burberry, The Perfume Shop, B&M, Curry's, PC World, and many more being affected by the COVID-19 outbreak are being forced to shut down indefinitely. The UK government has announced a wide-ranging draconian economic package to help support UK public services, businesses, and households. At his first budget, just weeks after becoming Chancellor of the Exchequer, Rishi Sunak announced £5bn of additional funding for public services, predominantly for the NHS. At the time, he stressed that more money would be made available if it was needed.

The government's main actions have focused on providing support to businesses and people whose incomes are affected by COVID-19 and the associated economic shutdown. This support includes tax cuts and public spending totalling around £55bn. These are permanent giveaways that will add to public borrowing and debt. The most expensive elements are exempting certain businesses in badly affected sectors from paying business rates, the government's commitment to pay 80 percent of furloughed employees' wages, and the commitment to pay 80 percent of self-employed people's past average profits, is simply unprecedented.

Deferrals of tax payments totalling over £40bn. These are not tax giveaways, the tax is still due, but the government has allowed businesses and people more time to make these payments, deferring them by around six months. Loan guarantees of up to £330bn. Loans will be issued by the British Business Bank (for smaller businesses) or the Bank of England (for larger businesses), with the government guaranteeing 80 percent of the value of the loans. Should a business fail, the government will pay 80 percent of the outstanding loan. But if businesses in receipt of these loans do not fail, these guarantees will not cost any money. The ultimate cost to the Exchequer is likely to be far below £330bn.

Overall, the measures that directly and immediately affect public sector borrowing (support for public services and tax cuts and public spending targeted at businesses and people) total around £60bn.

While most retailers have been forced to shut down to contain the pandemic, a raft of firms across the country, in sectors from manufacturing to fashion, are setting up emergency operations that often involve striking out into unfamiliar territory.

The most high-profile example has been the effort by industrial powerhouses to manufacture 30,000 medical ventilators, with household names such as Airbus, Dyson, Ford and Rolls-Royce, all pitching in with expertise and resources to make breathing apparatus and ventilators to help eradicate the disease, which is far away from their usual daily production. Even the fashion designer Ralph Lauren have also joined in to make designer masks and gloves so that people can protect themselves in style. Also, the banking group HSBC moved fast to help fund ventilator and PPE manufacturers by offering fast-track loan applications, cheaper interest rates and extended repayment terms.

Businesses across the UK have experienced a sharp fall in revenues because they are temporarily closed, have experienced a major loss of custom or had their supply chains disrupted. Many would face going under or having to lay off large numbers of staff unless they can get help to meet their running costs. This help is being offered in several ways. Using public spending to meet businesses' costs, the government has cancelled business rates payments for 2020/21 for businesses in England in the retail, hospitality, and leisure industries, and for nurseries. The Scottish and Welsh governments have also implemented this. Northern Ireland has announced a three-month holiday for all businesses from non-domestic rates (equivalent to business rates).

Even before the COVID-19 crisis, many businesses were due to receive a discount on their business rates in 2020/21. Those in the retail sector in England with a 'rateable value' between £15,000 and £51,000 (most such businesses) were already due to receive a 50 percent discount, as they did in 2019/20. To help further, the government is providing a cash grant worth £25,000 to these businesses and those in the hospitality and leisure sector within the same range of rateable values. Equivalent grants will also be provided by the Scottish and Welsh governments.

Businesses in the least valuable properties, those with a rateable value below £12,000 would pay no business rates at all (regardless of what sort of business they undertake). Those with a rateable value between £12,000 and £15,000 get a discount, due to the small business rates relief (SBRR). All businesses eligible for SBRR will now also receive a £10,000 cash grant.

A £10,000 grant will also be paid to all businesses eligible for rural relief, which covers low-value properties in sparsely populated areas. Equivalent grants will also be provided by the Scottish and Welsh governments.

Ordinarily, statutory sick pay (SSP) must be provided by employers to all employees earning above an average of £118 per week from the fourth consecutive day of illness and for up to 28 weeks. SSP is paid at a rate of £94.25 per week.

The government will reimburse small and medium size employers (SMEs; those with fewer than 250 employees) for up to 14 days of COVID-19-related sick pay per employee from the first day of absence. This includes periods when an employee is off work because they are self-isolating, even if they are not sick themselves. This is worth up to £188.50 per affected employee for qualifying employers.

On 20th March 2020, Mr. Sunak announced a new Coronavirus job-retention scheme, aimed at encouraging businesses to keep workers on their payroll, even if there is no work for them to carry out during the COVID-19 outbreak. The government will pay businesses up to 80 percent of the wage costs for workers who are furloughed. Payments can be backdated to 1st March 2020 and are capped at £2,500 a month for each worker.

This scheme will initially be in place for three months but will be extended if needed, and there is no limit to the amount of funding available. The Institute for Fiscal Studies has calculated that if 10 percent of employees were furloughed for three months, this would cost the government around £10bn. The government even introduced the emergency Coronavirus Bill which contains measures to prevent commercial tenants from being evicted if they fall behind on their rent between March and June.

The government has also taken action to help the low earners and those on benefits. Universal Credit payments will be increased by £1,000 a year for the next 12 months (with the same increase applied to the basic element of tax credits).

For those who are self-employed, the 'minimum income floor' (MIF) in Universal Credit will cease to apply for as long as the outbreak lasts. Ordinarily the MIF means that self-employed people are assumed to earn as much as they would in a full-time job paying the minimum wage, even if their income is lower than that. The chancellor has said that removing the MIF means that those who are self-employed will be entitled to receive an income from Universal Credit equal to the income from SSP that employees receive.

The government is also offering more generous support to meet rent payments for those who receive housing benefit, by increasing the housing allowance. Renters receiving housing benefit will now get an amount that is enough to cover the cheapest 30 percent of properties in their area.

The chancellor has also allocated £500m to a hardship fund, which will be used by local authorities to support vulnerable people. This will mainly be delivered in the form of reductions in the amount of council tax that people owe.

People who pay tax via self-assessment, mainly, but not exclusively, the self-employed, will also be given an extra six months to make their next payment for self-assessment income tax. The next payments, which were due at the end of July 2020, will now be deferred to the end of January 2021.

The government and the Financial Conduct Authority have also encouraged mortgage lenders to offer payment holidays for people who are in financial distress due to COVID-19. The government has encouraged landlords to do the same for renters. The emergency Coronavirus Bill includes provision for tenants to be given three months' grace before being evicted.

During the peak of the crisis, however, there will be higher demand for care services and staff will be stretched, as some will be unwell or self-isolating. The bill therefore suspends those duties, replacing them with a more general obligation to meet care needs where that is necessary, to avoid a breach of someone's human rights.

A similar story can be found in many provisions of the bill. The bill also relaxes existing legal requirements for discharging a patient from hospital, making decisions about patients with mental health conditions that mean they cannot make decisions for themselves, registering and certifying deaths and stillbirths, holding inquests, conducting cremations and managing dead bodies.

The government has limited resources and has to prioritise dealing with the massive crisis it now faces. At the same time, previous parliaments attached stringent requirements to these decisions because they are so important, and the consequences are so serious when they go wrong. For that reason, it is important that these suspensions only last so long as is necessary to respond to the pandemic. In trying to deal with that risk, we cannot avoid storing up others for the future.

CHAPTER 7

COVID-19 AND THE REST OF THE WORLD

Since it was recorded in late 2019 in China, the COVID-19 Coronavirus has spread around the world and has been declared a pandemic by the World Health Organization. By early spring, Europe had become the worst affected region, with Italy and Spain particularly hard hit. However, differences in testing mean that the number of cases may be understated for some countries.

The number of deaths is a more dependable indicator. The disease is hitting Italy and Spain with cruelty. But the trajectory in many countries is the same; the UK and US are a couple of weeks behind Italy in the progress of the epidemic.

The EU itself has introduced a strict border control. Travellers from outside are being turned away from airports and borders, after the 27-country block imposed a 30-day ban to halt the spread of Coronavirus, with Europe becoming the epicentre for the pandemic, with more than 150,000 confirmed cases of COVID-19 across the continent. German Chancellor Angela Merkel said: "Not since World War Two have we faced a challenge that depends so much on our collective solidarity."

As Italy struggles to deal with the rising number of dead, countries across Europe have brought in increasingly strict measures to ensure their citizens stay at home. Italy is now the worst affected country after China. As of 6th April 2020, Italy has 128,948 confirmed cases, with 15,887 deaths, which exceeds China's death toll, despite having fewer confirmed infections and a far smaller population. In the Northern province of Bergamo, the area hardest hit by the virus, the crematorium has started operating 24 hours a day. Funerals have been put on hold and churches are lined up with coffins as local morgues are full. Residents describe Bergamo as a ghostly place where only ambulances and hearses are on the road at night. Andrea Alessandro, a funeral home co-owner said, "Morgues and health institutions are collapsing. We were absolutely unprepared for an emergency of this kind."

In France, deaths from the virus had risen to 8,078, with a total of 70,478 confirmed cases and counting. Just like Italy, China and other countries, people in France are not allowed to leave their homes, unless it is for a sanctioned reason such as buying food, visiting a doctor, or going to work. Citizens must carry official paperwork stating why they are not at home, with fines of €135 (£119) for those caught breaking the rules.

Germany currently has the lowest mortality rate of the 10 countries most severely hit by the pandemic: the country has 100,123 confirmed cases, and 1,584 people have died. Experts say this is mainly because the outbreak started among younger people who tend to experience milder symptoms. Germany also started testing people with mild symptoms relatively early on. The government's central public health authority said the country has capacity for 160,000 tests per week.

Germany has closed schools and many businesses and public spaces. On Sunday, the government announced there would be no public gatherings of more than two people, people must always keep 1.5 meters (4.9 feet) from each other, and restaurants must close. German Chancellor Angela Merkel went into_quarantine on Sunday after being told she had come into contact with a doctor who tested positive for COVID-19.

According to the Spanish health minister, there are nearly 14,000 people dead in the country after testing positive for COVID-19. Confirmed cases topped 140,000, up from around 135,000 the day before.

Spain's health minister Jose Maria Sierra said that the uptick in deaths was for administrative reasons. He told a press conference: "What's happening is normal, and that there are oscillations after weekends where reporting is not so punctual." Spain has been one of the worst-affected countries by the coronavirus pandemic so far.

Only the US has had more confirmed cases, and only Italy has reported more virus-related deaths, according to Johns Hopkins University.

The Spanish Government ordered a lockdown several weeks ago to slow the spread of the virus, in a similar move to many other countries.

Prime Minister Pedro Sanchez said last week he would ask parliament to extend the quarantine until 26 April, arguing that the move was "saving lives".

The United States overtakes Italy to have the highest death toll from Coronavirus in the world. According to Johns Hopkins University, more than 20,000 people in the US have now died. The grim milestone comes shortly after the US became the first nation to record more than 2,000 Coronavirus deaths in a single day. With New York being the worst affected state in the country; Coronavirus has changed everything about life, and now it is upending the rituals of death. New Yorkers have been shocked by the grim scenes of ambulances constantly blaring down eerily deserted streets, body bags being forklifted into refrigerated trucks outside hospitals and now new trenches being dug on Hart's Island for possible mass burials. The remote cemetery, accessible only by boat, is a place regarded historically with sorrow because of its mass graves with no tombstones, just unclaimed bodies. The city's morgues can only handle so much before temporary burials for COVID-19 victims, once an absolute worst-case scenario, become necessary. Funeral directors talk openly about how scared and depressed the spiking death toll has left them. Even before this week's record number of deaths, some families have had to wait a week or more to bury and cremate their loved ones.

While the virus is slow to reach the African continent compared to other parts of the world, infection has grown exponentially in weeks and continues to spread, reaching the continent through travellers returning from hotspots in Asia, Europe, and the United States. Africa's first COVID-19 case was recorded in Egypt on 14th February. Since then, a total of 52 countries have reported cases. Initially, mainly confined to capital cities, a significant number of countries in Africa are now reporting cases in multiple provinces.

According to the World Economic Forum, it is the working class in the slums of a non-industrialized country, that would be the cook who transfers the infection to the traveller, who then brings the contagion to the industrialized nation. But we do not live-in stereotypical times. Think back to the Ebola outbreak in West Africa a few years ago. Seven out of the 28,646 suspected cases escaped the African continent in three and a half years, and yet the industrialized world was in a full-scale panic. Today, it is the industrialized world that is exporting an infectious disease to the global south. While maybe not as deadly as Ebola, COVID-19 is far more contagious.

With western countries, including the G7 nations, who are some of the wealthiest countries in the world, with all the modern medical technological equipment and expertise in health care that they have, also struggling to cope with the outbreak, it would be unmistakeable not to worry about African and Asian countries who are deemed third world and are regarded as some of the poorest countries in the world.

With the majority of these countries' health care systems drastically below par, with some countries not even having any sort of medical system in place, they have to rely on medical charity organisations like Médecins Sans Frontieres (Doctors without borders), Red Cross and others, while their leaders reap the wealth off the people and get away with it because of the corruption; how can they cope if COVID-19 trespasses into their territory since the virus knows no boundary. It is easy to see millions more dying because they do not even have the basics like PPE and if they do, most people do not have a clue how to use them. All of the countries in the western world like Britain, France, USA and so on, that would normally run to their aid, are also facing shortages of vital medical supplies like oxygen, masks, gloves, hand sanitizers and much more, and they are also facing all the tight restrictions and difficulties that come with the pandemic.

For instance, the United States has roughly 172,000 ventilators, and that is not enough for its 333 million population. Nearly all Western world countries that are affected have summoned the help of their manufacturers to make ventilators and PPE equipment as quickly as they can in order to help ease the pressure and save lives. Sierra Leone, a small country in West-Africa, has only thirteen ventilators to go round for its 7.98 million people. Also, the Central African Republic, with its 4.6 million population, only has three ventilators to go around the country. Liberia also has three and South Sudan has four for its large population.

Or take ICU beds. Some African countries have many. South Africa has 3,000. Somalia has only 15 ICU beds for the whole country. The largest city in the eastern Democratic Republic of Congo has perhaps two dozen ICU beds to serve a province with a population about the same size as Louisiana, with endemic Malaria, Malnutrition, Tuberculosis, and other diseases, that makes COVID-19 especially dangerous.

There are more than 12,300 confirmed cases of Coronavirus and 632 deaths across Africa thus far, and the numbers are rising daily. Several African countries are imposing a range of prevention and containment measures against the spread of the pandemic. According to the latest data by the John Hopkins University and Africa Centre for Disease Control on COVID-19 in Africa, the breakdown remains fluid as countries confirm cases as and when. The whole of Africa has rising cases with only two countries holding out as of 8th April.

In Liberia, the first case of COVID-19 came from someone who brought the disease from Switzerland. This traveller's household cook was the next in Liberia to test positive for the infection. Liberia's healthcare system was able to identify these first two cases and track down their contacts as a top-line system should. Many other African countries are also ahead of the curve, thanks to their experience with Ebola. In the Democratic Republic of Congo, Coronavirus has spread beyond the capital, Kinshasa, to the easternmost regions of the country, which until recently were still in the grip of an Ebola outbreak. In South Africa, which has the highest viral incidence on the continent, all provinces are now fighting the outbreak of COVID-19. Confirmed cases in Cameroon, Senegal and Burkina Faso are also widespread. While transmission rates are still low, the main fear is over what happens next.

In Asia, where the outbreak first began, the Chinese government moved swiftly to prevent the spread of the disease, instituting an unprecedented quarantine in Wuhan. Through a combination of high-tech scanning and tracking of its population, coupled with strict controls on people's ability to leave their homes, much less travel is made. Data can be faked; the collapse of a health care system due to an overwhelming number of patients is much harder to cover up. As it is now known, local and central authorities were not only slow to react to early reports of a mysterious new pneumonia-like illness in Wuhan, but even took steps to cover up the news. People issuing warnings on social media, most famously, Dr. Li Wenliang, who later died from the virus, were rounded up and reprimanded by the police. This let the pandemic spin out of control in the crucial early stages and arguably misled the rest of the world, including the World Health Organization, into downplaying the severity of the problem.

Hong Kong declared the new Coronavirus an emergency in late January, and its efforts since have managed to ward off a spike in cases. The public clearly still remembers the lessons from SARS and reverted to mask-wearing and social distancing protocols more easily than many other populations. Schools were closed and public gatherings banned, but businesses, including restaurants, have been allowed to stay open, albeit with strict protocols in place on distancing between people. In the eyes of many Hong Kongers, that success came because of the local government. As the rest of the world began to turn away travellers from China, the Hong

Kong administration dragged its feet on instituting border controls with mainland China, to the consternation of medical professionals. Even in mid-March, when Hong Kong instituted mandatory two-week quarantines on all foreign travellers, the rules did not apply to mainland China. Residents are also up in arms about shortages of face masks and other protective medical gear.

Indonesia, which is the world's fourth most populous country, now has the highest number of confirmed COVID-19 cases. Indonesia did not confirm its first case until 2nd March. Since then, cases have grown exponentially and on a daily basis, spreading to all 34 provinces across an archipelago of some 17,000 islands. The Health Ministry in Jakarta reports 5,923 positive cases following the country's largest daily jump of more than 400 new infections, along with 520 people dead, and yet, the light at the end of the tunnel is out of sight.

The first lockdown orders were not issued until over a month later, and the restrictions only applied to the Jakarta capital region and its population of some 30 million people. However, President Joko Widodo expanded the restrictions to some other parts of the country. He also advised the public to stay home during the Muslims' holy month of Ramadan, which starts later in April. This is by no doubt a big ask of the world's most populous Muslim country, as an estimated 19.5 million people travelled for the Eid al-Fitr holiday marking the end of Ramadan last year. In the region, only China has a higher death toll. Indonesia's government has faced criticism for not testing earlier and for not swiftly implementing strict social distancing and travel restrictions.

Japan managed to avoid the worst of the first wave of infections, to such an extent that up until mid-March officials were still talking about holding the Tokyo Olympics as scheduled. As of 1st April, Japan had only reported

2,500 cases and 60 deaths, many of those stemming from a single cruise ship. And that low number came despite businesses and borders remaining open at the time. Even as cases continued to climb in Tokyo, the central and local governments both declined to do much more than urge voluntary self-distancing practices, even after seeing the cautionary tales of lockdowns imposed too late in Italy and the United States. Prime Minister Shinzo Abe finally declared a state of emergency on 7th April, but even that was mostly voluntary and fell short of a true lockdown to the dismay of medical experts. Meanwhile, testing was limited, leading critics to say Japan probably had far more cases than its official count. Infections seem to be surging and Japan may have to play the same painful game of catch-up seen in Europe and the United States.

South Korea is perhaps the poster child for managing a COVID-19 outbreak. A combination of extensive testing, location tracking and contact tracing helped the country get control of the virus even after a massive surge in cases. Since then, countries around the world have been asking for advice and testing kits from South Korea, hoping to replicate its success. Though South Korea's response in the past month has been exemplary, the initial explosion of cases shows the extreme danger of noncompliance with social distancing protocols. The outbreak in Daegu was linked to a fringe religious group, the Sincheonji Church of Jesus, that urged supporters not to wear masks and to continue joining mass religious services, with attendees in the thousands. After one churchgoer brought COVID-19 into the mix, the outbreak quickly spiralled out of control. South Korea went from less than 30 cases on 13th February, when President Moon Jae-In declared the outbreak nearly over, to nearly 8,000 cases a month later.

After initially downplaying the pandemic, Cambodia has taken some belated steps including suspending foreign visas, declaring a state of emergency, and cancelling new year celebrations. With the country officially registering a few dozen cases as of now, it has also begun allocating more economic resources to the health sector amid concerns about the country's health system and its ability to handle rising cases. Hun Sen's initial public relations stunts, including offering to fly to Wuhan in China when it was the epicentre of the global coronavirus pandemic and allowing the Westerdam ship to dock in Cambodia, were reflective of the country's early downplaying of the virus, contrary to the advice of public health experts.

After some delay, the Philippines has enacted some restrictions, including cancelling fights, banning the entry of foreigners, and nixing key military exercises. The country also announced an economic stimulus package and social protection program. The Philippines' initial refusal to enact restrictions on travel and tourism from China has in part contributed to its continued status as one of the leading countries of reported COVID-19 cases in Southeast Asia. Testing is increasing but is still to be ramped up to adequate levels, and the government has been forced into measures such as stopping healthcare workers from going abroad to manage pressures on the health system at home. While not surprising, Duterte's rhetoric during COVID-19, with statements such as shooting lockdown violators on sight, has nonetheless complicated how the government's response is being viewed.

The Indian central government treated the pandemic with seriousness early on, taking measures, including a total ban on overseas arrivals to the country by air and ordering a nationwide lockdown. The 24th March lockdown order by

Indian Prime Minister Narendra Modi became the largest of its kind in the world, affecting the country's 1.3 billion people. Regional results have varied in India, with certain states like Kerala exhibiting a particularly robust response to the pandemic. New Delhi's rapid implementation of a national lockdown was done without clear communication, resulting in panic and frenzy as citizens initially wondered how to procure essential goods. Government communication had originally told Indians not to panic shop, resulting in a delayed panic after the lockdown order was announced. To make matters worse, law enforcement often acted overzealously in supporting lockdowns. India's COVID-19 testing capacity remains limited.

Tens of millions of Indians living below the poverty line and transient migrant rural-to-urban migrant workers were left with their livelihoods upended by the lockdown. With fiscal resources already under pressure, attempts by the central government to offer a social safety net are insufficient, with spending at under 1 percent of Indian GDP.

On 2nd April, the World Bank announced one of its largest emergency funding packages in South Asia for Sri Lanka, providing $128.6 million in financing to support emergency health systems and pandemic response. In March, Sri Lanka's election commissioner announced that the country's 2020 parliamentary election would be postponed indefinitely as a result of the pandemic. The election was scheduled originally for late April. Economically, Sri Lanka faces major supply chain concerns and has banned the export of essential goods, including food staples.

On 2nd March, Kazakh President Kassym-Jomart Tokayev cancelled the upcoming International Women's Day festivities. The holiday is a big deal in the former Soviet Union, and while Kazakhstan at that point had not confirmed any Coronavirus cases, its government made a proactive decision in the interest of public health. After confirming its first cases, Kazakhstan cancelled Nowruz celebrations in late March and even the scheduled 9th May Victory Day parade. The good news is Nur-Sultan has taken COVID-19 seriously and taken reasonable measures, including closing nonessential business, curtailing travel, and enforcing quarantines. Kazakhstan also has a government dashboard with the latest data and a hotline number.

As of early April, the novel Coronavirus has spread to every region in Kazakhstan with the highest number of cases present in Nur-Sultan and Almaty, the country's two largest cities, followed by the Karaganda region and Akmola region (just south of and surrounding the capital, respectively). The pandemic also brought with it an oil crisis as prices dropped with the implosion of global demand and Russia and Saudi Arabia held onto a nasty price war a little too long. Kazakhstan joined major oil producers to keep barrels off the market, cutting production as of 1st May. But it is probably too little too late, to avoid a bad hangover in oil-dependent economies like Kazakhstan from depressed prices. As the virus spreads in more rural areas, it will be more difficult for Kazakhstan to detect cases and care for patients. The differences in quality of healthcare infrastructure between Kazakhstan's urban and rural communities, and differences in access between the rich and the poor, could expose some ugly truths about social inequality in Kazakhstan.

The only good news about Turkmenistan under the current circumstances is how closed and isolated the country is, even in the best of times. With so few travellers normally, Turkmenistan might be spared by sheer virtue of its isolation from the outside world. Turkmenistan was in the midst of an economic crisis before the pandemic; its only lucrative business is gas and Ashgabat has essentially one customer: Beijing. Even if China is through the COVID-19 woods (and even if Turkmenistan miraculously escapes infection entirely), analysts expect lower gas demands in China as a by-product of slowing global demand for goods manufactured in the country. That will invariably be bad news for Turkmenistan.

Turkmenistan's leadership has built itself up on the premise that everything is golden and good in the land of the Turkmen thanks to Arkadag, the protector: President Gurbanguly Berdimuhamedov. With a cult of personality branded with his virility and sportsmanship, admitting there is an invisible foe that cannot be cured with a brisk bike ride chips away at the veneer. The ugly truth is that a lie that protects Berdimuhamedov's position is always preferable to the government over the truth that it is not equipped to handle this crisis. Turkmenistan may not have banned the word Coronavirus, but it sure has not used it enough.

CHAPTER 8

THE SECOND WAVE

Just when the world thought that they had some respite from the dangerous disease and were beginning to return to normal life, the Coronavirus disease struck again and this time, with a lot more vengeance than the first wave. People in Britain and other countries have just enjoyed a couple of weeks of freedom and all of a sudden, COVID-19 starts to spiral out of control again with confirmed cases rising daily like before, and deaths climbing also. Children had just managed to return to school, and universities just started teaching their students using combined teaching methods of face-to-face and online learning. Employees were getting used to working more from their offices again as commanded by the government, social distancing was being eased, sports and other event organisers were looking forward to welcoming fans and patrons back to their games, and pubs and clubs were running again, with only a few customers allowed in at one time, without the risk of breaking the social distance rule. Then more fear struck, and a second lockdown seems inevitable as the bleak of winter lingers around the corner.

Some countries such as Albania, Bulgaria, Czech Republic, Montenegro, North Macedonia are seeing higher case numbers in August than they did earlier in the year. France, the UK, Poland, the Netherlands, and Spain are likely dealing with the much-feared second wave and have started taking action to curb it. France, for example, declared 16,096 new cases and the Netherlands 2,541 in a day; the highest figures they have recorded so far. There are trends that may explain the deterioration. The surge comes just after the summer vacation season. The World Health Organization has suggested the increase could be partly down to the relaxation of measures and people dropping their guard, and evidence indicates young people are driving the second surge in Europe. Despite the rising numbers of cases and recent deaths in Europe, the continent still compares favourably to the United States. Europe has reported 4.4 million cases and 217,278 deaths among a population of 750 million, while the US has reported 6.7 million cases and 198,000 deaths in a population of 330 million. Boris Johnson told reporters that the UK is "now seeing a second wave coming in" and that it was inevitable. "Obviously, we are looking very carefully at the spread of the pandemic as it evolves over the last few days. There is no question, as I have said for weeks now, that we could (and) are now seeing a second wave coming in. We are seeing it in France, in Spain, across Europe."

"Several countries are reporting more daily COVID-19 cases than they did during the first wave in March, though the higher numbers may be due to more people being tested now. In some member states, the situation is now even worse than in the peak during March," said Stella Kyriakides, the European Union's commissioner for health and food safety, on 24th September 2020. "This is a real cause for concern."

Researchers at Imperial College London said recently that the number of cases had been doubling roughly every 7.7 days in England, and that the reproduction rate was as high as 1.7. If the virus continues to spread at that rate, the

UK will see about 10,000 new cases a day in the next two weeks, with 300 to 400 hospital admissions a day. In April, about 3,000 people were admitted to hospital every day, and the number of deaths doubled every four days. At the current rate of infections and admissions, it would take five weeks to reach those levels.

Prime Minister Boris Johnson has outlined a range of new lockdown measures to fight the so-called "second wave" of COVID-19 infections taking place across the U.K, measures he says could last for six months. Mr. Johnson told the country that pubs and hospitality venues would have to close at 10pm and that people should work from home if they can. Johnson also restricts all gatherings to a maximum of six people, except for weddings and funerals, where they are allowed to exceed the number to fifteen for weddings and thirty people for funerals. It marks a sharp turn from previous government advice, which encouraged workers to return to offices again as the government tries to help the economy recover. Johnson said that the UK was at a "perilous turning point" and that the government had to act. Liverpool, Warrington, Hartlepool, Newcastle, and Middlesbrough, in a bid to curb the spread of coronavirus are all placed under local lockdown. The Prime Minister also introduced a £10,000 fine for anyone who repeatedly breaks the rules. Businesses are also legally required to take customers' contact details so they can be traced if there is an outbreak. They can also be fined up to £10,000 if they take reservations of more than six people, or do not enforce social distancing. Staff in hospitality venues must now wear masks along with customers when they are not seated at their table to eat or drink. The penalty for not wearing one or breaking the "Rule of six"2 has doubled to £200 for a first offence.

In Scotland, First Minister Nicola Sturgeon said, "That Scotland will impose a 10pm curfew on pubs and restaurants."

She also announced a ban on meeting inside other people's homes. Up to six people from two different households can meet outdoors, including in private gardens. However, there will be no limit on the number of children under 12 who can meet or play together outdoors.

Young people aged 12 to 18 are exempt from the two-household limit and can meet outdoors in groups of up to six. Ms Sturgeon raised the possibility of a two-week "Circuit breaker", with further restrictions for Scotland in October, although she said no decision had been made. She urged people not to book foreign trips during the October holiday.

More than three million people in Madrid have had new restrictions imposed on their lives as Spain tries to control the most serious second wave of COVID-19 infections in Europe. From this weekend, 2nd October 2020, people can travel outside their home districts for essential journeys only. Bars and restaurants cannot serve after 22:00. And a maximum of six people is permitted to meet in any setting. The measures have been demanded by Spain's federal government. They also take effect in nine towns around the capital. The reason for these strict measures in Spain is because signs of the second wave of COVID-19 infections now breaking over Spain can be seen at the emergency admission unit of the 12 de Octubre Hospital, one of the biggest in Madrid. Every hour, ambulances arrive with new patients. Some of the sick are helped into wheelchairs; others, already needing oxygen, have to be stretchered in by medical staff wearing full protective gear.

A red warning signal indicating the seriousness of Spain's predicament is that, at many hospitals across Madrid, existing intensive care units (ICU) are once again full with COVID-19 patients. Hospitals are having to use overflow capacity prepared at the height of the pandemic, including beds usually reserved for burns patients and for post-operative recovery. La Paz Hospital, another of Madrid's biggest, all 30 ICU critical care beds are occupied. "There are more patients than we can attend in critical care units," says Juan José Río, Medical Director at La Paz. "Psychologically it's the worst thing because all the staff here are afraid that the Tsunami will come again."

In the United States, President Donald Trump announced on Twitter at 1am that he and (FLOTUS), the First Lady of the United States, Melania Trump, have tested positive for Coronavirus and are now self-isolating on 2nd October 2020. Mr Trump, aged 74 and therefore in a high-risk group, wrote on Twitter": "We will get through this together." The president has "mild symptoms" of coronavirus, White House officials say. Mr Trump's announcement comes just over a month before the presidential elections on 3rd November 2020, where he faces Democratic challenger Joe Biden. The development came after Hope Hicks, one of Mr Trump's closest aides, tested positive. The stunning development after months of debilitating losses, set against a badly mismanaged federal response overseen by a commander-in-chief who repeatedly downplayed the crisis injected new turmoil into the country's leadership at a moment of deep national strain.

Trump was flown by Marine One to Walter Reed National Military Medical Centre on the afternoon of Friday 2nd October 2020, and will remain there for several days, the White House said. He was administered a dose of Regeneron and remains fatigued, according to a memorandum from his physician. "As of this afternoon the President remains fatigued but in good spirits. He's being evaluated by a team of experts, and together we'll be making recommendations to the President and First Lady in regard to next best steps," Navy Commander Dr. Sean Conley wrote. This drug,

Regeneron, was sadly not available to millions of Americans, which is why the death and infection rates soared like they did, but it was available all of a sudden for one so-called important man, POTUS (President of the United States). Can you imagine? The taxpayers who are paying for these so-called leaders to rule us, well, the leaders are having all the best drugs, all the best medical experts around them 24 hours a day, and first-class medical care, while the taxpayers get the basics of it. So unfair.

The majority of Trump's circle also tested positive for COVID-19. The Republican National Committee Chairwoman Ronna McDaniel received a positive test. Senator Mike Lee, a Utah Republican, and Notre Dame President Fr. John Jenkins both announced on Friday they had tested positive for coronavirus.

Both Lee and Jenkins were at the White House on Saturday when Trump announced his Supreme Court nominee, Amy Coney Barrett. Barrett tested negative after being diagnosed with Coronavirus late this summer and was recovering. Democratic presidential nominee Joe Biden tested negative after sharing a debate stage with Trump.

Both Ivanka Trump and Jared Kushner, senior advisers to Trump, tested negative for coronavirus Friday morning, the White House said. Barron Trump, the President and first lady's 14 years old son, also tested negative, according to Stephanie Grisham, Melania Trump's chief of staff.

Vice President Mike Pence's press secretary confirmed on Twitter that Pence and his wife, Karen, had tested negative for COVID-19. Pence's doctor released a letter saying that the vice president, "Is not considered to have had close contact with any individuals who have tested positive for COVID-19," and he "does not need to quarantine". President Trump discharged himself two days later, saying he is immune to the virus and that he is well now. The President said that he has an election campaign to run, and he cannot run it from the Walter Reed Medical Centre.

Fast forward to 3rd November 2020, Americans elect their 46th president of the United States, President-Elect: Joseph Robinette Biden Jr and their first black and Asian female Vice President-Elect: Kamala Harris in the midst of the pandemic. However, the incumbent president Trump is refusing to concede and is calling the election a fraud after Joseph Biden clearly defeated him with 273 electoral votes to 214. This win for President-Elect Biden came mainly because President Donald Trump has always downplayed the virus, and blocked certain funding needed for health professionals to save lives, leading to over 200,000 deaths in the United States alone and counting. Rather than conceding, President Trump is filing lawsuits in various states for election fraud and is said to be preparing himself for a second term.

CHAPTER 9

THE VACCINES AND THEIR PROBLEMS

Within less than twelve months after the beginning of the COVID-19 pandemic, several research teams around the world rose to the challenge and developed vaccines that can supposedly protect people from the SARS-CoV-2, the virus that causes COVID-19. This is a historic move as it usually takes up to ten years or more to research, develop and try a vaccine. With striking speed, the United Kingdom became the first country to approve a COVID-19 vaccine that has been tested in a large clinical trial. On 2nd December 2020, UK regulators granted emergency-use authorization to a vaccine from the drug firms Pfizer and BioNTech, just seven months after the start of clinical trials. The Pfizer/BioNTech COVID-19 vaccine began being rolled out in the UK on 8th December 2020. Since then, millions of people have been vaccinated, including the elderly, care workers and frontline NHS staff.

A UK grandmother became the first person in the world to be given the Pfizer COVID-19 jab as part of a mass vaccination programme. Margaret Keenan, who will turn 91 soon, said the injection she received at 06:31 GMT was the “Best early birthday present”. At University Hospital,

Coventry, UK, matron May Parsons administered the injection to Ms Keenan. "I feel so privileged to be the first person vaccinated against COVID-19," said Ms Keenan, who is originally from Enniskillen, Co Fermanagh. "My advice to anyone offered the vaccine is to take it. If I can have it at 90, then you can have it too," she added.

COVID-19 vaccines work by using a harmless version or component of the SARS-CoV-2 Coronavirus to train the immune system so that when the immune system encounters the virus for real, it can fight it off. Around four in ten people who received their first dose had at least one after effect in their arm, most commonly pain and swelling a day or two after the jab. Most symptoms happen in the first two days after vaccination, with headache, fatigue and chills or shivers being the most common side effects. After effects are more common the second time around, with around one in five people who received their second dose of the Pfizer vaccine reporting at least one systemic effect. Similarly, more people experienced effects in their arm after their second dose, with about half reporting local symptoms like pain and swelling in the injected area. Some of the common side effects of the coronavirus vaccine may include tenderness, swelling or redness at the injection site, headache, muscle ache, feeling tired and fever (temperature above 37.8°C). A less common side effect is swelling of the glands. This starts a few days after being inoculated with the vaccine and it may last for up to two weeks. This is to be expected and is a sign of the immune system responding to the vaccine.

There is also the COVID-19 Moderna vaccine with a 94.1 percent efficacy, which works by preparing the body to defend itself against COVID-19. It contains a molecule called mRNA (Messenger Ribonucleic Acid) which has instructions for making the spike protein. This is a protein on the surface of the SARS-CoV-2 virus which the virus needs to enter the body's cells. When a person is given the vaccine, some of their cells will read the mRNA instructions and temporarily produce the spike protein. The person's immune system will then recognise this protein as foreign and produce antibodies and activate T cells (white blood cells) to attack it. If the person comes into contact with the SARS-CoV-2 virus, their immune system will recognise it and be ready to defend the body against it. The mRNA from the vaccine does not stay in the body but is broken down shortly after vaccination. Messenger Ribonucleic Acid plays a vital role in human biology, specifically in a process known as protein synthesis. MRNA is a single-stranded molecule that carries genetic code from DNA in a cell's nucleus to ribosomes, the cell's protein-making machinery.

The Moderna vaccine has the same mild side effects as the Pfizer and BioNTech vaccine but with more serious side effects such as redness, hives and a rash at the injection site; a rash occurred in less than 1 in 10 people. Itching at the injection site occurred in less than 1 in 100 people. Swelling of the face, which may affect people who had facial cosmetic injections in the past, and weakness in muscles on one side of face (acute peripheral facial paralysis or palsy) occurred rarely, in less than 1 in 1,000 people. Allergic reactions have occurred in people receiving the vaccine, including a very small number of cases of severe allergic reactions (anaphylaxis). As for all vaccines, the COVID-19 Moderna vaccine should be given under close supervision with appropriate medical treatment available.

Moderna's vaccine, which was developed in collaboration with the US National Institute of Allergy and Infectious Diseases, works in the same way as the one produced by Pfizer and BioNTech. Both consist of RNA molecules encased in lipid nanoparticles. The RNA in both vaccines encodes a slightly modified form of the SARS-CoV-2 protein known as spike, which enables the virus to infect human cells.

Since the Pfizer vaccine was rolled out in the United Kingdom and the United States, there have been occasional reports of recipients experiencing severe allergic reactions called anaphylaxis after being injected. There have been no signs of such reactions so far in the Moderna trial, which excluded people who have had anaphylactic reactions to previous vaccines, but not those with other allergies, such as reactions to food. The two vaccines differ in the composition of the lipid nanoparticle that encases the RNA, and Moderna's formulation allows the vaccine to be stored at a higher temperature than Pfizer's, which must be kept at −70 °C, much colder than a normal freezer. Moderna's vaccine can be stored in a −20 °C freezer for 6 months, and in a refrigerator (at about 4 °C) for 30 days. This promises to streamline the logistics of deploying the vaccine, particularly in rural areas and in countries with limited health-care infrastructure.

On the other hand, the Oxford AstraZeneca vaccine also had a successful trial. Its Phase III clinical trials in the UK, Brazil, and South Africa, confirmed that the AstraZeneca vaccine is safe and effective at supposedly preventing COVID-19, with no severe cases and no hospitalisations, more than 22 days after the first dose. Results demonstrated vaccine efficacy of 76 percent after a first dose, with protection maintained to the second dose. With an inter-dose interval of 12 weeks or more, vaccine efficacy increased to 82 percent.

There is also the Johnson and Johnson vaccine which is set to be a cost-effective alternative to the Pfizer and Moderna vaccines and can be stored in a refrigerator instead of a freezer. This vaccine only requires people to be inoculated once with a single dose, unlike its competitors that require two shot dosages. Trials found it prevented serious illness but was 66 percent effective overall when moderate cases were included. The vaccine was developed mainly by J&J's Janssen Pharmaceutical division. The company has agreed to provide the US with 100 million doses by the end of June. The first doses could become available to the US public. The UK has ordered 30 million doses, the EU ordered 200 million, and Canada ordered 38 million doses, while 500 million doses have also been ordered through the COVAX scheme to supply poorer nations.

Can the COVID-19 vaccine be a threat to younger people who wish to have children in the future? Well, some people think that some if not all the COVID-19 vaccines can affect your reproductive system which will prevent you from having children later in life. Could this be the governments' agenda to reduce the population of the world in order to reduce their spending on state benefits, housing and much more, by causing so many people to die with the virus and by creating infertility in the masses so that more children won't be conceived, or is this another conspiracy theory to put fear into people in order to prevent them from taking the vaccine?

Amid a handful of reports of blood clots which can block several arteries including the artery that drains blood from the brain, a growing list of countries, primarily in Europe, have suspended the use of the AstraZeneca vaccine as the continent faces a third wave of COVID-19. The move plunges an already slow European vaccination drive into further disarray. Austria was the first country to sound the alarm over potential blood clots caused by the AstraZeneca vaccine and Denmark became the first country to suspend the jab.

Denmark, Norway, Austria, Estonia, Latvia, Lithuania, Ireland, and Luxembourg, have halted some or all of their AstraZeneca COVID-19 vaccinations. Italy banned the use of a specific AstraZeneca batch of the vaccine, which is believed to be the ABV2856, adding that it may consider additional measures, if needed, in coordination with the European Medicines Agency (EMA). The Italian agency stressed, however, that at present there is no proven connection between the inoculation of the vaccine and the adverse reactions. Austria also suspended the use of a batch of the AstraZeneca vaccine on Sunday after a 49-year-old woman died as a result of blood clots, 10 days after receiving the jab.

Also, several governments outside Europe outlined similar plans. So far, Thailand, Indonesia, the Democratic Republic of the Congo, and Venezuela have either halted rollouts of the vaccine or announced plans to suspend inoculations. AstraZeneca has strongly defended the vaccine, arguing that there is no increased risk of fatal brain haemorrhages and blood clots. That stance has largely been backed up by experts who have said that instances of blood clots and rarer thrombocytopenia cases are no higher among those who received the jab than the general population.

Responding to the decision taken by Denmark, the vaccines safety lead of the UK's Medicines and Healthcare Products Regulatory Agency (MHRA), Dr. Phil Bryan, said that the action undertaken by Danish authorities was a precautionary measure and that, as of yet, it has been unconfirmed that the blood clot was a result of the Oxford vaccine. "Blood clots can occur naturally and are not uncommon. More than 11 million doses of the COVID-19 AstraZeneca vaccine have now been administered across the UK safely," the doctor says. The UK would later go on to eat its words and follow other countries in taking precautionary measures. The UK has suspended the Oxford AstraZeneca vaccine for under 30s and offered them other alternatives like the Pfizer or Moderna vaccines after several people in the UK also started having blood clots. The Johnson and Johnson vaccine is the only vaccine that has been ordered by the courts to put a clear warning of possible blood clots.

More than 23 million people in the UK and counting have received at least one dose of a coronavirus vaccine, one of the biggest inoculation programmes the country has ever launched. In a race against a faster-spreading variant of the virus, ministers have pinned their hopes of easing a third national lockdown by vaccinating as many adults as possible by the summer. Since vaccinations began in the United States on 14th December 2020, more than 100 million doses have been administered, reaching 19.3 percent of the total US population, according to federal data collected by the Centers for Disease Control and Prevention. The US is currently administering over 2.2 million shots a day. On the 19th March 2021, President Biden reached his election campaign promise of inoculating one hundred million Americans in his first 100 days as president, and achieved this goal within 58 days of his presidency. India has delivered 26.2 million doses of the vaccine to its one billion population and counting, followed by China, which has delivered 52 million doses so far to its one billion people. Worldwide, the vaccination effort continues with a current dosage of 335 million administered so far around the world and counting. On the same day as President Biden's election campaign promise record, Prime Minister Boris Johnson received his first COVID jab of the Oxford and AstraZeneca vaccine at London's St. Thomas' Hospital, where he was admitted to Intensive Care the year before with the Coronavirus disease.

It is said that Botswana has secured enough vaccines to inoculate its adult population against the Coronavirus disease, according to Mosepele Mosepele, the deputy coordinator of the presidential COVID-19 task team.

According to Bloomberg Quint, the diamond-rich southern African country agreed deals with vaccine manufacturers to receive enough vaccines to inoculate 1.9 million people out of a total population of 2.4 million. Those eligible for the vaccination, or are above 18 years of age, are 1.6 million people. Among the shipments, doses to cover 1.1 million people will come from Johnson & Johnson, and 250,000 from Moderna.

US regulators approved the Pfizer and BioNTech vaccine for 12 to 15-year-olds on Monday 10th May 2021. According to the Guardian Newspaper, this will widen the country's inoculation program, even though vaccination rates have slowed significantly. The vaccine has been available under an emergency use authorization (EUA) to people as young as 16 in the United States. Today's decision means that the FDA is amending the EUA to include children aged 12 to 15. The vaccine makers said they had started the process for full approval for those ages last week.

Prime Minister Boris Johnson said on the 6th June 2021 that he will urge G7 leaders of wealthy nations to commit to vaccinating the whole world against the Coronavirus by the end of 2022. According to BBC News, the PM said he is asking his fellow leaders to "Rise to the greatest challenge of the post-war era by vaccinating the world by the end of next year. I'm calling on my fellow G7 leaders to join us to end this terrible pandemic and pledge we will never allow the devastation wreaked by the Coronavirus to happen again," he said.

The US, France, Germany, Italy, and Japan, have all said how many doses they will donate to the global vaccine-sharing programme COVAX, but the UK and Canada are yet to outline figures on their planned contributions. However, during the G7 meeting in Cornwall, England, including the

UK, agreed to donate one billion vaccines to the scheme to help poorer nations like Kenya, South Africa and many more.

On 5th June 2021, UK regulator MHRA (Medicines and Healthcare products Regulatory Agency) approved the use of the Pfizer-BioNTech vaccine in children aged 12-15, saying it is safe and effective in this age group and the benefits outweigh any risks.

On the 16th June 2021, Health Secretary Matt Hancock says he will make the vaccine mandatory for all Care home workers who are dealing with vulnerable people. He also said that any member of staff who is not vaccinated will be redistributed elsewhere in the NHS or face being sacked. Many members of staff have voiced their anger with the government for trying to force the vaccine on them, and many have made it clear that they would rather be sacked than take the vaccine. However, Care organisations have warned against compulsory vaccination as it could drive away good care workers.

Countries like the Seychelles, Bahrain, and Mongolia, who have vaccinated the majority of their populations are reporting high number of Coronavirus cases. This rise in cases is believed to have happened due to the ineffectiveness of the Chinese vaccine. These countries chose this vaccine as it was easily accessible to them to tackle COVID-19. In the Seychelles, Chile, Bahrain, and Mongolia, about 50 percent to 68 percent of the populations have been fully inoculated with Chinese vaccines, outpacing the United States. They are also among the top ten countries with the worst COVID-19 outbreaks. China's Sinopharm vaccine has an efficacy rate of 78.1 percent and the Sinovac vaccine has an efficacy rate of 51 percent.

Moreover, the Chinese companies have not released much clinical data to show how their vaccines work at preventing transmission. According to Business Standard, William Schaffner, who is a Medical Director of the National Foundation for Infectious Diseases at Vanderbilt University, said the efficacy rates of Chinese shots could be low enough "To sustain some transmission, as well as create illness of a substantial amount in the highly vaccinated population, even though it keeps people largely out of the hospital."

Also, there is a report that 43 percent of people who have been fully vaccinated in England were amongst the 117 people that died from the Delta variant on the 1st June 2021. Just what is going on here? Are the vaccines really working as they should or is the Delta variant just too deceptive? Are we really protected by the vaccines against COVID-19?

The EU has now rolled out its vaccine passport for all 27 member states and says those who were jabbed with the Indian made AstraZeneca Covishield vaccine can enter the continent as long as they were jabbed in the UK by the NHS. The European Commission says that it is the decision of its member states to let in those who were vaccinated by the Covishield vaccines or not. Countries like Greece are already accepting these vaccines. For instance, they are accepting China's Sinovac vaccine and Russia's Sputnik and many others. However, the UK's NHS COVID Pass has not been approved by the EU. In essence, anyone with a COVID passport will be exempted from testing or quarantining when travelling through European borders.

CHAPTER 10

THE OUT OF CONTROL SECOND WAVE AND THE NEW VARIANTS

On Monday 4th January 2021, the UK Prime Minister, Boris Johnson said: "Since the pandemic began last year, the whole of the United Kingdom has been engaged in a great national effort to fight COVID-19, and our efforts would have worked if it wasn't for a new COVID-19 variant that has emerged, and it is frustrating and alarming to see the speed at which it is growing." Mr Johnson added that our hospitals are under more pressure than they have been at any time during the pandemic, and that 80,000 people in the country tested positive for COVID-19 on 29th December. He then announced that England will adopt a national lockdown, saying: "In England, we must therefore go into a national lockdown tough enough to contain this variant." This third lockdown in England came weeks after Wales, Northern Ireland, and various other countries all went into their third lockdown. This spike happened after the government eased the lockdown restrictions for the Christmas and New Year holidays, which caused lots of people to forget about the pandemic and they started travelling freely and mixing with friends and families for celebrations.

So, what are these new variants and why the need for a third lockdown? Researchers in the UK have concluded that the new strain is between 29 percent and 91 percent more likely to kill infected Brits. The London School of Hygiene and Tropical Medicine said the new variant could be 1.35 times more deadly; Imperial College London said it was between 1.36 or 1.29 more deadly, and the University of Exeter found it may be 1.91 times more deadly. Whichever one of these medical institutions got their figures correct, it just shows that the new strain is very deadly and more contagious. According to BBC Health, all viruses, including the one that causes COVID-19, mutate. These tiny genetic changes happen as the virus makes new copies of itself to spread and thrive. Most are inconsequential, and a few can even be harmful to the virus's survival, but some can make it more infectious or threatening to the host humans. There are now many thousands of different versions, or variants, of the virus circulating, but experts are worried about a small number of these.

In the United Kingdom, a new variant called Kent B.1.1.7 has emerged with an unusually large number of mutations. This variant spreads more easily and quickly than other variants. This variant was first detected in September 2020 and is now highly prevalent in London and most parts of the country. India and many other countries have stopped operating airlines to and from the UK due to the discovery of this new variant B117 virus or VUI2020/21. The travel ban is likely to last until the end of December 2020. It has since been detected in numerous countries around the world, including the United States and Canada. In South Africa, another variant called 1.351 has emerged independently of the variant detected in the UK. This variant, originally detected in early October, shares some mutations with the variant detected in the UK. There have been cases caused by this variant outside of South Africa. In Brazil, a variant called

P.1 emerged and was identified in four travellers from Brazil, who were tested during routine screening at Haneda airport outside Tokyo, Japan. This variant contains a set of additional mutations that may affect its ability to be recognized by antibodies. Both the new South African and UK "Kent" variants appear to be more contagious, which is a problem because tougher restrictions on society may be needed to control the spread. Dr Simon Clarke, who is an expert in cell microbiology at the University of Reading, said: "The South African variant has a number of additional mutations including changes to some of the virus' spike protein which are concerning."

The spike protein is what Coronavirus uses to gain entry into human cells. It is also the section that vaccines are designed around, which is why experts are worried about these mutations. Professor Francois Balloux, from University College London, said: "The E484K mutation has been shown to reduce antibody recognition. As such, it helps the virus SARS-CoV-2 to bypass immune protection provided by prior infection or vaccination." These new variants are highly responsible for the high death rates since the pandemic started and because of that, the government now must impose tougher restrictions on its people. Restrictions like the stay-at-home order unless you are going for an essential journey like restocking food or medications or going for an important medical appointment. Restrictions like the closure of all schools and colleges apart from those children whose parents are key workers. Remote learning is introduced. All non-essential shops must once again close their doors to their customers. People cannot leave their homes except for certain reasons, like in the first lockdown. These include essential medical needs, food shopping, daily exercise, and work for those who cannot do so from home. Early years places such as nurseries are also suspended, and university students are told not to return to campuses, but to continue their courses online instead.

Elsewhere in the UK, Scotland's First Minister Nicola Sturgeon also announced a new lockdown across mainland Scotland, Skye, Arran, Bute and Gigha for the rest of January. Level three restrictions apply in the remaining areas of the nation.

Wales has been in a national lockdown since 20th December 2020, and schools and colleges in the nation will remain closed until at least 18th January 2021.

In Northern Ireland - which entered a six-week lockdown on 26th December 2020 – "stay at home" restrictions will be brought back into law from Friday.

The most feared South African variant of COVID-19 has now made its way to the UK, causing the British government to impose even tougher measures.

It has been reported that the new Indian variant is so strong that the newly developed vaccines could be ineffective against it. The variant has the ability of evading the immune system. The new variant is a first of its kind triple mutant variant since COVID-19 began, which just shows that the virus has evolved and gotten stronger. This variant has caused death tolls to rise significantly in India, and hospitals are nearing their capacities. As of 23rd April 2021, at least six hospitals are at full capacity.

The variant which is called B.1.617 was only confirmed on 25th March 2021 by the Indian government, and now scientists and doctors are growing concerned that this particular variant could be more transmissible and deadlier than all its predecessors. A total of 182 cases of this variant have been detected in the UK. There were 162 confirmed cases in the five weeks up to 16th April 2021, forcing Boris Johnson to postpone his trip to India and adding it to the UK's travel red list corridor. The Indian variant has the E484Q mutation, which is very similar to the one found in the South African and Brazil variants, and it has the V382L in its spike. This is a sub-lineage of B.1.617, and it also has the L452R mutation found in the Californian variant. India's cases have gone up dramatically since the middle of March 2021, after weeks of steady decline.

New Delhi is now undergoing another week of strict lockdown measures from 19th April 2021, which is due to be extended by Prime Minister Narendra Modi. India set the world record for having the highest daily COVID-19 cases of 332,730 within a 24-hour period, and 2,263 deaths reported within the same 24-hour period. These are recurring daily figures and are said to be higher than what is currently being announced by the government.

Hospitals have padlocked their gates as they have run out of beds and oxygen supplies, which is causing many people who could have survived the virus to die, and those who are fighting to save their loved ones from dying from COVID-19 are being forced to scout for oxygen supplies three hundred miles, putting their lives at risk of catching the virus by queuing up for oxygen gas across the country. Even when they get the oxygen cylinder, unknowing to them, the cylinder is empty and does not have any oxygen gas in it. They have to go back to fill the cylinder up with oxygen gas, which requires more money, and when they finally return to their loved ones, these people do not have the necessary medical equipment and expertise to connect the oxygen to their patients. This second wave has descended on India in the harshest manner; the United Kingdom, France, the United States and others, who are also struggling with the pandemic, have sent in aid like oxygen, ventilators and many other medical supplies to help India recover from their deadly wave.

Crematoriums across the country are inundated with bodies waiting to face this old-age Indian ritual. One of New Delhi's residents, Nitish Kumar, was forced to keep his dead mother's body at home for nearly two days while he searched for space in the city's crematoriums, a sign of the deluge of death in India's capital where coronavirus cases are surging. Kumar finally cremated his mother, who died of COVID-19, in a makeshift mass cremation facility in a parking lot adjoining a crematorium in Seemapuri in northeast Delhi. "I ran pillar to post but every crematorium had some reason. One said it had run out of wood," said Kumar, wearing a mask and squinting his eyes that were stinging from the smoke blowing from the burning pyres. According to one resident, "Children who were 5 years old, 15 years old, and adults 25 years old or more are being cremated. Newlyweds are being cremated. It's difficult to watch."

With a current population of over 1 billion, and poverty levels extremely high, and with more than 800 million people considered poor in India, bad hygiene conditions are the cause of diseases such as Cholera, Typhus and Dysentery, where children especially suffer and die. Poverty in India impacts children, families, and individuals in a variety of different ways, through high infant mortality and multiple people sharing the same toilet. This is said to have been the reason why the Coronavirus has blown out of control, with 20 million or more confirmed cases present in the country as of 3rd May 2021.

Cyrus and Adar Poonawalla are the founder and CEO of the Serum Institute of India. It is the world's largest vaccine-producing company in the world's largest vaccine-producing nation. Serum makes vaccines for Measles, Tetanus, Diphtheria,

Hepatitis, and many other diseases. It specializes in generic versions, exports to 170 countries, and estimates that two-thirds of the world's children are inoculated with its vaccines. The Serum company produced the Covishield which was developed by AstraZeneca and Oxford University. There is also another company called Bharat Biotech, which is a Hyderabad based pharmaceutical company responsible for the production of the Covaxin. Why then is India struggling to acquire enough COVID vaccine for its people?

According to dw.com, the World Trade Organization (WTO) has announced that a proposal by India and South Africa to temporarily suspend intellectual property (IP) rules related to COVID-19 vaccines and treatments has hit a roadblock after wealthy countries backed down on the idea. The two developing countries say the IP waiver will allow drug makers in poor countries to start production of effective vaccines sooner.

India and South Africa approached the global trade body in October, calling on it to waiver parts of the agreement on Trade-Related Aspects of Intellectual Property Rights (TRIPS Agreement). The suspension of rights such as patents, industrial designs, copyright, and protection of undisclosed information, would ensure timely access to affordable medical products, including vaccines and medicines, or scaling-up of research, development, manufacturing and supply of medical products essential to combat COVID-19.

The proposal was vehemently opposed by wealthy nations like the US and Britain, as well as the European Union, who said that a ban would stifle innovation at pharmaceutical companies by robbing them of the incentive to make huge investments in research and development. Is this how humanity is now? Valuing money over life. However, on 6th May 2021, President Biden's administration supported the idea to waiver the intellectual property in order to save lives. Will other rich nations follow the US? The state of Mumbai and other states across the country have halted their inoculation progress due to the shortage of the Coronavirus vaccine.

This Indian variant is now threatening the UK's freedom to ease the COVID-19 restrictions fully. Everyone in the UK is longing for this freedom where they are free to go out as they wish to visit family and friends, indoor dining and pubs, and other businesses once they become fully operational, so that life can return to normal. This variant is set to beat the record of the Kent variant in many ways. Dubbed as the Delta variant, the Indian variant is now considered to be the dominant strain of Coronavirus in the UK, Australia, and sixty more countries worldwide. On 14th May 2021, it was reported that cases of the Indian variant jumped up from 560 to over 1,200 confirmed cases in the UK, with Bolton being the hotspot this time around, along with four confirmed deaths. On 6th June 2021, the UK's Health Secretary Matt Hancock said that the Indian variant, now known as the Delta variant, is 40 percent more transmissible than the UK's Kent Alpha variant, and 70 percent of COVID cases in the UK are of this deadly variant. It has also been said that the Delta variant seems to infect younger people a lot, which has caused President Biden during his visit to the UK for the 2021 G7 meeting to urge younger people in the country to take the jab.

If care is not taken, the Indian variant could get even worse, especially with the large number of protests which are filled with thousands of people or more, going on in London and around the world because of the new Israeli-Palestinian conflict. These protests can fuel the Indian variant and cause it to increase the death toll of the disease and grind the world to a halt once again through another lockdown. Boris Johnson has brought forward second vaccination jabs from twelve weeks to eight weeks in order to tackle this Indian variant.

The Rotterdam Eurovision 2021 song contest is in limbo, with there being the possibility of it being cancelled after a member of the Polish delegation tested positive for COVID-19 during a test run for the show. The Polish delegate is not the first COVID-19 case at Eurovision 2021. Although the organisers have managed to keep the artists and press as safe as they could, a handful of people had been found to be positive. All positive cases were barred entry to the venue and have been told to go into self-isolation. The Dutch city was due to host the contest in 2020 before the event was cancelled due to the ongoing COVID-19 pandemic. And with Holland reporting 5,000 cases a day of the virus, will it be cancelled, or will the organisers risk it all?

In Brazil, cases of COVID-19 are rising as the more transmissible P1 variant spreads across the country. In April, nearly one third of all daily COVID-19 fatalities in the world were in Brazil, although Brazil makes up only 2.7 percent of the world population. On 2nd April 2021, there were 12.8 million cases and over 325,000 deaths. In the week 21–27 March, there was a daily 0.8 percent rise in cases and a 1.9 percent rise in deaths; lethality has risen from 2 percent to 3.3 percent since late 2020. The new variants circulating in Brazil have become a serious cause of concern to neighbouring countries.

On 29th March 2021, 17 out of the 27 federal states reached adult-ICU-bed occupation rates of 90 percent or more; among the 27 capital cities, 21 displayed the same rates, and seven of them had reached their full capacity or were working above it. In most points of care, the number of available beds, although insufficient, results from successive expansions due to the high demand. Despite these efforts, by 25th March, 6,371 people were waiting for an ICU-bed. On the very same day, 496 people lost their lives while on the waiting list for an ICU bed in the state of Sao Paulo alone.

Brazil has the second highest death toll in the world from COVID-19, and experts are warning that a current surge in cases may not peak for several weeks. The rapid spread of a coronavirus variant first discovered in Brazil has been a major cause for concern around the world. The Brazilian public health institute Fiocruz says it has detected 92 variants of coronavirus in the country.

In particular, the P1 variant has become a cause for concern because it is thought to be much more contagious than the original strain and has been spreading across Latin America and the world. This new variant is suspected to have emerged in the city of Manaus in the Brazilian Amazon in November 2020, before it caused hospitals there to collapse in January 2021, and spread to the rest of the country. Scientists believe the current vaccines should still work against the Brazilian variant, although perhaps not quite as well, and new variants could emerge in the future that are different again.

Dr Nicolelis says: “Brazil is not only the epicentre of the pandemic worldwide, it is a threat to the entire effort of the international community to control the pandemic. We are brewing new variants every week.”

Hospitals in Brazil are being forced to intubate Coronavirus patients without sedatives amid critical medicine shortages caused by the country’s current outbreak. A doctor at the Albert Schweitzer municipal hospital in Rio de Janeiro, said that doctors are resorting to tying patients to their beds in order to ventilate them. According to Sky News, Brazil has a daily death rate averaging around 3,000 a day, with oxygen supplies for intensive care patients being at breaking point. If care is not taken, and proper measures are not implemented as soon as possible, the Brazilian pandemic could end up being as bad, or even worse, than the Indian variant.

So, what has made the pandemic so worse in Brazil? According to nature.com, researchers are devastated by the recent surge in cases and say that the government's failure to follow science-based guidance in responding to the pandemic has made the crisis much worse. They add that President Jair Bolsonaro's administration has publicly undermined the scientific guidance, while refusing to implement protective national lockdowns and spreading misinformation.

"Being a scientist in Brazil is so sad and frustrating," says Jesem Orellana, an epidemiologist at the Oswaldo Cruz Foundation's centre in Manaus. "Half of our deaths were preventable. It's a total disaster." Since Brazil's vaccination campaign was rolled out, only 6 percent of the population has been vaccinated, which means they are ahead of some South American countries but are falling behind Europe and North America. Vaccinations have not been proven to slow the virus transmission, which could be accelerated by the new variant known as P1. This variant is said to reinfect people who have had COVID-19 before, which proves that people cannot really be immune from the virus as scientists initially thought.

Along with Brazil and India, Peru has recorded more than 62,126 deaths among its population of 33 million. The Peruvian government reported that 47,000 Peruvians have died from COVID-19 so far, though excess death figures suggest that, because of undertesting, the actual total is closer to 85,000.

Also, Japan, which was supposed to host the Tokyo 2020 Olympic Games, which was postponed due to the pandemic and was scheduled to start on 23rd July 2021, might have to postpone the games again after several states across the country have declared a state of emergency.

According to the latest Japanese government figures, there were 7,521 new cases on 12th May 2021, including 969 infections in the host-city Tokyo. Regions scheduled to host athletes have been hard hit, including the eastern region of Chiba, where the US track and field team had been due to have a training camp. According to Sky News, Hokkaido, which is hosting test events for the Olympic marathon, reported 1,029 cases on Wednesday 12th May 2021.

Some athletes are also questioning whether the Games should go ahead, with tennis stars Rafael Nadal, Serena Williams and Naomi Osaka raising their concerns. Nadal said he was unsure what his calendar will look like this summer, while Williams's doubts stem from the possibility of not being able to travel with her three-year-old daughter Olympia.

Japan's world number two Osaka said that rising COVID-19 levels in Tokyo are a "Big cause of concern" and said she was not sure if the Games should go ahead.

CHAPTER 11

A THIRD DEADLY WAVE

The much feared third wave of the Coronavirus SARS-CoV-2 has arrived on planet Earth, starting in India where the deadly, most contagious Delta variant emerged from and has now made itself well known around the world, from South Africa to Oman to Russia. Countries like Russia, who have been boasting all along about their successful handling of the pandemic, have now introduced new strict COVID-19 restrictions across the Kremlin state, which includes regional lockdowns and compulsory vaccinations in their belated response to a third Coronavirus wave sweeping the country. Russia has seen an explosion of new infections since mid-June, driven by the Delta variant. The surge comes as officials in Moscow are pushing vaccine-sceptic Russians to get inoculated after lifting most anti-virus restrictions in late 2020. Russia has confirmed 5,451,291 cases of Coronavirus and 133,282 deaths since the pandemic began. Russia's total excess fatality count since the start of the coronavirus pandemic is around 475,000.

According to Moscow Times, on Monday 28th June 2021, Russia confirmed 21,650 new Coronavirus cases, the highest number since January, and 611 deaths. From Monday, all restaurant and cafes in Moscow will require patrons to present an official QR code confirming their vaccination status, immunity, or negative PCR test results. All Russian travellers, vaccinated or not, will have to present a negative PCR or rapid test result on arrival in Greece, Athens. Travelers from Russia will also have to undergo another test as soon as they arrive.

South African authorities have also re-imposed COVID-19 restrictions in a last-ditch effort to stem the sharp rise in COVID-19 cases that is ravaging the country's economic heartland. The wave of infections has been driven by the spread of the more transmissible Delta variant, poor countermeasures imposed by the government, and public fatigue with existing restrictions of the pandemic, as well as a shortage of vaccines in the country. President Cyril Ramaphosa said that all gatherings, indoors and outdoors, would be banned for 14 days, along with the sale of alcohol, dining in restaurants and travel to or from the worst-hit areas of the country. An extended curfew would also be imposed, and schools will shut early for the holidays.

South Africa's rising cases are part of a resurgence across Africa, with a peak expected to exceed that of earlier waves, as the continent's 54 countries struggle to vaccinate even a small percentage of their populations. The World Health Organization (WHO) has repeatedly appealed for vaccines for Africa, saying a fast-surging COVID third wave is outpacing efforts to protect populations, "Leaving more and more dangerously exposed".

According to the Guardian Newspaper, South Africa's Gauteng province, which is the most populous part of the country, COVID-19 patients are waiting for hours, even days on stretchers in A&E wards before being allocated a bed, officials said. Unlike past waves, this time around, the hospital system was not coping, said Dr Angelique Coetzee, the Chair of the South African Medical Association. All leisure travel in and out of Gauteng province, which now accounts for about 60 percent of the country's new cases, will be banned. Visits to nursing homes and other congregant settings will be restricted. Schools and other educational institutions will close for the winter holidays early. The restrictions will remain in place for 14 days, at which time they will be re-evaluated, the President said.

Cases are also rising quickly in 12 other countries in the continent, although health systems are already pushed to breaking point in many more African countries. In Namibia, Uganda, and Zambia, among other places, they are running out of oxygen and hospital beds are full. The WHO calculates that, within weeks, the Africa-wide caseload of the third wave will surpass the peak of the second, which in turn was higher than the first.

Despite a successful vaccination campaign in many countries, Europe is facing an overwhelming third wave of COVID-19 infections. According to Worldometers, Europe has recorded 46,790,150 cases of the Coronavirus disease, out of which, 1,076,623 have lost their lives. On 4th June, it recorded over 19,925 more cases, with 751 people having died and 50,000 recovered from the respiratory illness.

According to Republic World, France is the worst affected in the continent, having the highest number of positive cases of 5,694,076 and 109,857 deaths. The country, famous for its galleries and museums, has announced it will open its tourist sites for vaccinated foreign visitors only. It also held a pilot outdoor concert recently.

The UK, which is leading the continent in COVID-19 vaccinations, has reported 4,499,878 cases and 127,812 deaths. Prime Minister Boris Johnson has downplayed the threat of emerging B.1.617 Delta variant cases. On 26th June

2021, there was a 72 percent rise in cases in England, and yet Prime Minister Johnson is sticking to his plans to lift all COVID-19 restrictions across England. Other countries which are facing a surging infection include Italy, which has recorded 4,225,163 cases and 126, 342 fatalities, and Germany, which has registered 3,701,690 cases and 89,605 deaths. Spain, which is gearing up the summer, is also witnessing a rise in the cases.

The Sultanate of Oman in the Arabian Peninsula is also experiencing its third wave of the Coronavirus outbreak, according to His Excellency Mohammed Al Hosni, Undersecretary of Ministry of Health, during an interview with Oman TV: “We are at the beginning of the third wave, and we must do everything to reduce these infections.” Seventy percent of Oman’s population will be vaccinated by end of the year. Thirty percent of vaccinations will be completed by end of June, says Al Hosni.

The Ministry of Health reported 728 new cases of COVID-19, taking the tally of infections in the sultanate to 151,528. It also reported seven coronavirus-related deaths, increasing the number of mortalities to 1,629. Seventy-two people were admitted to hospital during the past 24 hours, taking the number of inpatients to 356, including 104 in intensive care.

According to BBC News, seven Australian cities are now in lockdown as authorities scramble to prevent the spread of the highly contagious Delta coronavirus variant. Officials reported a case rise to more than 200 cases. Nearly half the population, more than 12 million people are under stay-at-home orders in Sydney, Brisbane, Perth, Darwin, Townsville, and the Gold Coast. The outback town of Alice Springs also entered a snap lockdown after cases emerged in South Australia.

Authorities fear the virus could now spread to nearby Aboriginal communities which are already vulnerable. Across the country, state leaders said they were facing a “pressure cooker situation” as new cases emerged. Many leaders have urged faster vaccinations as just 5 percent of the 25,788,215 population is fully vaccinated. The Australian government has been slow to roll out the vaccines and they are also sending out mixed messages to the people saying that only over 60s are vulnerable to the virus and they should take the jab. However, anyone under 40 who wishes to be vaccinated with the AstraZeneca jab must consult their GP before taking the vaccine. This shows the public that their government does not trust the vaccines, which puts doubt into people’s mind.

CHAPTER 12

TRACK AND TRACE AND PRIVACY CONCERNS

What is the track and trace app and how does it work? A contact tracing app is designed to let people know if they have been in close contact with someone who later reports positive for COVID-19.

It could pinpoint exactly who needs to be in quarantine and who doesn't, making it key to easing up social distancing measures. The purpose of the contact-tracing app is to try and track down people and alert them of the need to self-isolate faster than traditional methods.

Users who download the app to their phone can voluntarily opt-in to record details of their symptoms when they start to feel unwell. The app keeps a trace of others who have been in close contact through Bluetooth signals that transmit an anonymous ID. These low energy Bluetooth signals perform a digital 'handshake' when two users come into close contact but keep the data anonymous. The app will track your every move and create an electronic ID, which you will not have any control over.

India's Aarogya Setu COVID-19 tracing app was initially rolled out in early April as being voluntary, but it is now starting to be pushed by various government, non-government and private agencies as being mandatory. The move has resulted in several instances where the absence or the presence of the app has been used to make people behave in a certain way, and the government's poor record on the pervasive use of digital tools has left many experts sceptical about the use of data from the app to build a health surveillance system or curb civil liberties.

The app is quickly becoming an integral part to people being able to return to normal life. In one instance, in the North Indian territory of Chandigarh, 190 people who violated a curfew imposed due to the lockdown were forced to download the app before being released from police detention. In another, a photographer was denied entry into a pharmacy inside a residential complex in Noida after refusing to download the app. Residents of Noida have challenged the mandatory push as an executive overreach. Note that India is not alone in this mandatory use of the track and trace app. China also launched several types of apps that use either direct geo-localisation via cell phone networks, or data compiled from train and airline travel or highway checkpoints. Their use was systematic and compulsory, and it played a key role in allowing Beijing to lift the lockdown and halt contagions, with no new deaths reported since mid-April. South Korea, for its part, issued mass mobile phone alerts announcing locations visited by infected patients, and ordered a tracking app installed on the phone of anyone ordered into isolation; these aggressive measures have been credited with helping curtail the outbreak.

Also, many governments are looking for ways out of the restrictive physical distancing measures imposed to control the spread of Severe Acute Respiratory Syndrome Coronavirus 2 (SARS-CoV-2). Some governments have suggested, including Chile, Germany, Italy, the UK and the

USA, the use of Immunity Passports, which could be either a digital or a physical document that certify whether an individual has been infected and is purportedly immune to SARS-CoV-2. Individuals in possession of an Immunity Passport could be exempt from physical restrictions and could return to work, school, and daily life. Even though this COVID-19 Immunity Passport has not been implemented yet because the World Health Organization says that "There is currently no evidence that people who have recovered from COVID-19 and have antibodies are protected from a second infection. At this point in the pandemic, there is not enough evidence about the effectiveness of antibody-mediated immunity to guarantee the accuracy of an Immunity Passport." This does not mean that the Immunity Passport will not be implemented at a later date.

President Donald Trump, along with the Department of Defence and the US Department of Health and Human Services, announced a $138 million contract with ApiJect Systems America for 'Project Jumpstart' and 'RAPID USA', which together will dramatically expand US production capability for domestically manufactured, medical grade injection devices, beginning in October 2020. Spearheaded by the Department of Defence's Joint Acquisition Task Force (JATF), in coordination with the HHS Office of the Assistant Secretary for Preparedness and Response, the contract will support 'Jumpstart' to create a US based, high-speed supply chain for prefilled syringes, beginning later this year, by using well established Blow-Fill-Seal (BFS) aseptic plastics manufacturing technology, suitable for combatting COVID-19 when a safe and proven vaccine becomes available.

According to Apiject, the syringe is a technological advancement as it will enable Health officials who are running a scheduled vaccination program or an urgent pandemic response campaign to make better decisions if they know when and where each injection occurs. With an optional RFID/NFC tag under the label on each BFS prefilled syringe, Apiject will make this possible. Before giving an injection, the healthcare worker will be able to launch a free mobile app and "tap" the NFC tag on the prefilled syringe's label to their phone. The app will capture the tag's unique dose-level serial number, GPS location and date or time, then upload the data to a government database. Aggregated injection data provides health administrators with an evolving real-time 'injection map'.

The UK's NHS Track and Trace programme, which is a centralised system, has been under-performing, leading to many infected people going undetected. Most hospitality venues in England and Wales were previously running QR code-based systems to collect customer details, which every user of a smartphone with QR scanning functionality was able to use. Customers would scan these codes and input their names and mobile phone numbers, so that if a positive case of COVID-19 was traced to the venue they could be informed by human track and trace workers. However, there are numerous reports of problems with the system. Many people have complained about breach of their privacy while others said that the service is giving them false results. The app told them to self-isolate because they have come into contact with someone who has tested positive, but when they check, this was incorrect. The NHS COVID-19 app was also briefly unable to log the results of tests carried out in Public Health England's labs, NHS hospitals and from many other test venues, all of which created a huge problem for people as they were unable to get accurate, or any, test results whatsoever.

CHAPTER 13

A NEW GLOBAL FINANCIAL CRISIS

COVID-19 is not only a global pandemic and public health crisis, it has also severely affected the global economy and financial markets. As disease outbreaks are not likely to disappear in the near future, proactive international actions are required to not only save lives but also protect economic prosperity. Significant economic impact has already occurred across the globe due to reduced productivity, loss of life, business closures, trade disruption and decimation of the tourism industry. The spread of Coronavirus across the world is picking up speed and it is already clear that there has been an impact on global supply chains in many industries that are dependent on supplies from China. Some automotive manufacturers have already had to suspend production due to a lack of parts. Given the spread of COVID-19, a similar scenario can be expected in Europe; as well as production interruptions, interference in the operation of catering establishments and shopping malls, a drop in tourism revenues and the start of working from home have all occurred. Its spread has also left businesses around the world counting costs and wondering what the recovery could look like. Big shifts in stock markets, where shares in companies are bought and sold, can affect the value of pensions or individual savings accounts (ISAs). The FTSE, Dow Jones Industrial Average, and the Nikkei, all saw huge losses as the number of COVID-19 cases grew.

Many people have lost their jobs or seen their incomes cut due to the coronavirus crisis. Unemployment rates have increased across major economies as a result. In the United States, the proportion of people out of work has hit 10.4 percent, according to the International Monetary Fund (IMF), signalling an end to a decade of expansion for one of the world's largest economies. Millions of workers have also been put on government-supported job retention schemes as parts of the economy, such as tourism or hospitality, came to a standstill under lockdown. However, the data differs between countries. France, Germany, and Italy, provide figures on applications, for example, whereas the UK counts workers currently enrolled on the scheme.

If the economy is growing, that generally means more wealth and more new jobs. It is measured by looking at the percentage change in gross domestic product (GDP), or the value of goods and services produced, typically over three months or a year. But the IMF says that the global economy will shrink by 3 percent this year. It described the decline as the worst since the Great Depression of the 1930s.

According to the House of Commons Library, it is said that "The magnitude of the recession caused by the Coronavirus outbreak is unprecedented in modern times. UK GDP (Gross Domestic Product) was 26% lower during the depth of the crisis in April than it was only two months earlier in February. While a recovery is underway, there is uncertainty over how fast economic activity will regain lost ground. Economic prospects depend greatly on how the COVID-19 caseload evolves. The possibility of a resurgence in cases presents the greatest risk to the economic outlook. The reactions of consumers and businesses to the uncertainty will also play an important role in the speed of the recovery. Consumers may be reluctant to return to 'normal' spending patterns. This may be due to health concerns but also perhaps due to concerns over their income. A key factor will be how high unemployment levels rise. Particularly important is how many employees currently furloughed will return to work and how many will become unemployed. Uncertainty may also dampen businesses' inclination to invest."

The coronavirus outbreak is significantly affecting public finances. The UK government's budget deficit is increasing as tax revenues fall and government spending increases. Government debt is, therefore, increasing. The measures the UK Government has taken to support businesses, workers and household incomes may cost over £190 billion this year. The longer the crisis continues, the cost to the government will continue to rise. The budget deficit in 2020/21 is likely to reach a level last seen during World War II. The outlook for public finances in the coming years depends on the strength of the economic recovery. As the economy recovers the Government's deficit will decrease. Tax receipts will recover and providing support to individuals, workers and businesses will fall. The extent to which the economy recovers will depend on how much permanent damage or scarring there has been.

In addition to the substantial burden on healthcare systems, COVID-19 has had major economic consequences for the affected countries. The COVID-19 pandemic has had a direct impact on income due to premature deaths, workplace absenteeism and a reduction in productivity, all of which has created a negative supply shock, with manufacturing activity slowing down due to global supply chain disruptions and closures of factories. For example, in China, the production index in February declined by more than 54 percent from the preceding month's value. In addition to the impact on productive economic activities, consumers typically changed their spending behaviour, mainly due to decreased income and household finances, as well as the fear and panic that accompany the epidemic. Service industries such as tourism, hospitality, and transportation, have suffered significant losses due to a reduction in travel. The International Air Transport Association projects a loss in airline revenue solely from passenger carriage of up to $314 billion. Restaurants and bars, travel and transportation, entertainment and sensitive manufacturing are among the sectors in the US that have been the worst affected by the COVID-19 quarantine measures. The unemployment rate in the U.S. has reached a record level of 11 percent for the week ending 11th April 2020.

Furthermore, the economic impact of the COVID-19 pandemic will be heterogeneous across country's income distribution. For example, office workers are more likely to transition to flexible working arrangements during restrictions, while many industrial, tourism, retail and transport workers will suffer a significant reduction in work due to community restrictions and low demand for their goods and services.

According to the World Bank, the June 2020 Global Economic Prospects describes both the immediate and near-term outlook for the impact of the pandemic and the long-term damage it has created to prospects for growth. The baseline forecast envisions a 5.2 percent contraction in global GDP in 2020, using market exchange rate, determines the deepest global recession in decades, despite the extraordinary efforts of governments to counter the downturn with fiscal and monetary policy support. The crisis highlights the need for urgent action to cushion the pandemic's health and economic consequences, protect vulnerable populations, and start the long-lasting recovery process. Every region is subject to substantial growth downgrades. East Asia and the Pacific will grow by a scant 0.5 percent. South Asia will contract by 2.7 percent, Sub-Saharan Africa by 2.8 percent, Middle East and North Africa by 4.2 percent, Europe and Central Asia by 4.7 percent, and Latin America by 7.2 percent. These downturns are expected to reverse years of progress toward development goals and tip tens of millions of people back into extreme poverty.

Emerging market and developing economies will be buffeted by economic headwinds from multiple quarters: pressure on weak health care systems, loss of trade and tourism, dwindling remittances, subdued capital flows and tight financial conditions, amid mounting debt. Exporters of energy or industrial commodities will be particularly hard hit. Demand for metals and transport-related commodities such as rubber and platinum used for vehicle parts has also tumbled. While agriculture markets are well supplied globally, trade restrictions and supply chain disruptions could yet raise food security issues in some places.

Global growth is projected to rise from an estimated 2.9 percent in 2019 to 3.3 percent in 2020 and 3.4 percent in 2021, a downward revision of a 0.1 percentage point for 2019 and 2020 and 0.2 for 2021, compared to those in the October World Economic Outlook (WEO). According to IMF, in advanced economies, growth is projected to stabilize at 1.6 percent in 2020–21 (0.1 percentage point lower than in the October WEO for 2020, mostly due to downward revisions for the United States, Euro area and the United Kingdom, and downgrades to other advanced economies in Asia, notably Hong Kong SAR following protests).

In the United States, growth is expected to moderate from 2.3 percent in 2019 to 2 percent in 2020 and decline further to 1.7 percent in 2021 (0.1 percentage point lower for 2020 compared to the October WEO). This moderation reflects a return to a neutral fiscal stance and anticipated waning support from further loosening of financial conditions.

Growth in the Euro area is projected to pick up from 1.2 percent in 2019 to 1.3 percent in 2020 (a downward revision of 0.1 percentage point) and 1.4 percent in 2021. Projected improvements in external demand support the anticipated firming of growth. The October 2019 WEO projections for France and Italy remain unchanged, but the projections have been marked down for 2020 in Germany, where manufacturing activity remains in a contractionary territory in late 2019, and for Spain, due to carryover from stronger-than-expected deceleration in domestic demand and exports in 2019.

In the United Kingdom, growth is expected to stabilize at 1.4 percent in 2020 and firm up to 1.5 percent in 2021, unchanged from the October WEO. The growth forecast assumes an orderly exit from the European Union at the end of January, followed by a gradual transition to a new economic relationship.

Japan's growth rate is projected to moderate from an estimated 1 percent in 2019 to 0.7 percent in 2020 (0.1 and 0.2 percentage point higher than in the October WEO). The upward revision to estimated 2019 growth reflects healthy private consumption, supported in part by government countermeasures that accompanied the October increase in the consumption tax rate, robust capital expenditure, and historical revisions to national accounts. The upgrade to the 2020 growth forecast reflects the anticipated boost from the December 2019 stimulus measures. Growth is expected to decline to 0.5 percent (close to potential) in 2021, as the impact of fiscal stimulus fades.

For the emerging market and developing economy group, growth is expected to increase to 4.4 percent in 2020 and 4.6 percent in 2021 (0.2 percentage point lower for both years than in the October WEO) from an estimated 3.7 percent in 2019. The growth profile for the group reflects a combination of projected recovery from deep downturns for stressed and underperforming emerging market economies, and an ongoing structural slowdown in China.

Growth in emerging and developing Asia is forecast to inch up slightly from 5.6 percent in 2019 to 5.8 percent in 2020 and 5.9 percent in 2021 (0.2 and 0.3 percentage point lower for 2019 and 2020 compared to the October WEO). The growth markdown largely reflects a downward revision to India's projection, where domestic demand has slowed more sharply than expected amid stress in the nonbank financial sector and a decline in credit growth. India's growth is estimated at 4.8 percent in 2019, projected to improve to 5.8 percent in 2020 and 6.5 percent in 2021 (1.2 and 0.9 percentage point lower than in the October WEO), supported by monetary and fiscal stimulus as well as subdued oil prices. Growth in China is projected to inch down from an estimated 6.1 percent in 2019 to 6.0 percent in 2020 and 5.8 percent in 2021. The envisaged partial rollback of past tariffs and pause in additional tariff hikes as part of a "Phase One" trade deal with the United States is likely to alleviate near-term cyclical weakness, resulting in a 0.2 percentage point upgrade to China's 2020 growth forecast, relative to the October WEO. However, unresolved disputes on broader US-China economic relations as well as much needed domestic financial regulatory strengthening are expected to continue weighing on activity. After slowing to 4.7 percent in

2019, growth in ASEAN-5 countries is projected to remain stable in 2020 before picking up in 2021. Growth prospects have been revised down slightly for Indonesia and Thailand, where continued weakness in exports is also weighing on domestic demand.

Growth in emerging and developing Europe is expected to strengthen to around 2.5 percent in 2020–21 from 1.8 percent in 2019 (0.1 percentage point higher for 2020 than in the October WEO). The improvement reflects continued robust growth in central and eastern Europe, a pickup in activity in Russia, and ongoing recovery in Turkey, as financing conditions turn less restrictive.

In Latin America, growth is projected to recover from an estimated 0.1 percent in 2019 to 1.6 percent in 2020 and 2.3 percent in 2021 (0.2 and 0.1 percentage point weaker respectively than in the October WEO). The revisions are due to a downgrade to Mexico's growth prospects in 2020–21, due to continued weak investment, as well as a sizable markdown in the growth forecast for Chile, affected by social unrest. These revisions are partially offset by an upward revision to the 2020 forecast for Brazil, owing to improved sentiment following the approval of pension reform and the fading of supply disruptions in the mining sector.

Growth in the Middle East and Central Asia region is expected at 2.8 percent in 2020 (0.1 percentage point lower than in the October WEO), firming up to 3.2 percent in 2021. The downgrade for 2020 mostly reflects a downward revision to Saudi Arabia's projection on expected weaker oil output growth following the OPEC decision in December to extend supply cuts. Prospects for several economies remain subdued owing to rising geopolitical tensions (Iran), social unrest (including in Iraq and Lebanon), and civil strife (Libya, Syria, Yemen).

In sub-Saharan Africa, growth is expected to strengthen to 3.5 percent in 2020–21 (from 3.3 percent in 2019). The projection is 0.1 percentage point lower than in the October WEO for 2020 and 0.2 percentage point weaker for 2021. This reflects downward revisions for South Africa (where structural constraints and deteriorating public finances are holding back business confidence and private investment), and for Ethiopia (where public sector consolidation, needed to contain debt vulnerabilities, is expected to weigh on growth).

CHAPTER 14

WHEN PANDEMICS COLLIDE: COVID-19 AND OBESITY

Childhood obesity and the Coronavirus are pandemics that negatively affect the health and well-being of children. The disease of childhood obesity has risen to pandemic levels. Childhood obesity continues to be a significant public health problem across the world and can be legitimately described as a global pandemic. It is estimated that by 2020, 158 million children and adolescents between the ages of 5–19 would be categorized as obese.

With financial stress increasing in families, since many people who are the bread winner for their family have lost their jobs or have been furloughed, along with the lack of physical exercise, as gyms are closed and children are locked indoors weeks upon end without going outside, has meant that food consumption has increased, with majority of the food being unhealthy because most families are unable to access heathy foods due to shortages or poverty. Also, the lack of healthy school meals since schools are closed and children's sleeping patterns having changed, has caused childhood obesity to worsen. All children do is eat a lot and watch the screens, whether it be watching a favourite movie or show, or playing video games on the TV, or on the tablet, or using computers for doing their online Zoom or Teams lessons.

In the UK, 9.7 percent and 20.2 percent of children aged 4 and 10–11 years respectively, were classed as obese or severely obese in 2018/19, based on data from the National Child Measurement Programme (NCMP). Compared to 2009/10 NCMP data, this represents an overall increase in trend. There has been a 2 percent increase in obesity, which is really striking. The study looked at a large paediatric primary care network and found the number of patients with obesity had increased from 13.7 percent to 15.4 percent in the US.

One parent said, "We thought we were feeding her correctly. She was getting fruits, vegetables, home-cooked meals regularly, but I think our issue was, we kind of let her have treats like chocolates and sweets. To be told the news that she was obese, it was horrible. We were very upset as we just could not understand how this is possible."

The 2019–2020 school year ended abruptly due to the pandemic, in order to stem the flow of infection. With face-to-face teaching in classrooms scraped, and children forced to stay at home, their learning moved online. Some countries, including the UK, went into complete lockdowns; this meant that families were confined in their homes. People were only allowed to go out for essential shopping and for a once a day restricted physical activity.

The lockdown led to people panic buying and storing long shelf-life high calorie processed foods to minimize their trips to the supermarket. Lockdowns are stressful periods, even for children, having a potential impact on their behavioural attitude. According to the Journal of Diabetes and Metabolic Disorders, "As a result of the lockdown and the stress it brings even for children, this likely influenced their attitude and leads them to binge eating of high calorie food and sugary drinks." Such actions increase the chances of weight gain, especially if the amount of daily physical activity was reduced due to the restrictions in place. However, some households used the lockdown as the perfect opportunity to cook homemade food and buy more fresh food more than before. Unfortunately, a lot of people lost their jobs during the pandemic, leading to financial difficulties, this subsequently led people to buy less expensive shelf-stable food. Such a socioeconomic change might have a negative effect on what kind, and the amount, of food children eat.

School closures resulted in the absence of structured physical activity sessions, which can lead to a higher risk of extended inactive periods and a rise in weight gain among children. Children living in densely populated areas and in small apartments were faced with greater challenges due to limited space or opportunities for physical activity, and consequently, they stand a higher chance of gaining weight. Additionally, during the first wave of COVID-19, people were advised to stay at home, and playgrounds, non-essential shops, and leisure centres were closed.

Before the COVID-19 pandemic, it was reported that young people largely used online platforms to communicate with others, access social media and play video games. Because schools shifted to virtual learning during the pandemic, this resulted in higher screen time for kids. Although this was advantageous for teachers and their pupils' educational purpose, the increase in screen time can, to a great extent, worsen inactive habits, as well as increase the risk of anxiety, depression, and distraction. A link has been established between increased body mass index and the percentage of body fat as screen time increases. Furthermore, screen time is also associated with an increase in snacking and an eventual increase in weight.

Family environment is considered to have a huge impact on the everyday lives of children. According to the Journal of Diabetes and Metabolic Disorders, "If the family follows an inactive lifestyle, it is almost unavoidable that the child will stick to such habit." The presence of obesity during pregnancy has been linked with childhood obesity as well as diabetes and cardiovascular disease. At the start of the COVID-19 pandemic, pregnant women were said to be part of the vulnerable group and were even asked to stay at home by some governments. As restrictions were slowly lifted, pregnant women had to make the difficult decision of whether accessing antenatal care was more important than the risk of exposure to COVID-19. It is important to bear in mind that remote antenatal care was made available in some places.

However, these unprecedented times may have brought some families closer together. Parents working from home meant that they had more valuable time with their children. Nonetheless, parents faced new challenges in trying to balance taking care of their children, home-schooling them, while at the same time working from home. Stronger family bonds are expected to have been developed as more family events under one roof had to be catered for. The COVID-19 pandemic has made the situation on children's health, in terms of obesity, worse. However, there is a worrying link between childhood obesity, poor mental health, low self-esteem, and marginalisation. The July 2020 Obesity Strategy measures, which sets a goal for obesity in the UK, is inadequate. The policy is also inefficient, as it contains many strategies which have already been proposed but never implemented by the Westminster government.

Obesity in adults is defined by the World Health Organisation as a (BMI) Body Mass Index equal to or greater than 30. Overweight is a BMI equal to or greater than 25.

Most adults went into lockdown with a normal weight and came out of the lockdown/s obese or overweight. A report from the World Obesity Federation published on 4th March 2021, showed that death rates from Coronavirus have been ten times higher in countries where more than half of the population is obese. Prior to the outbreak of COVID-19, around 2.8 million people died per year because they were obese or overweight. According to the World Obesity Federation, the current pandemic might contribute to an increase in obesity rates as weight loss programmes, (which are often delivered in groups) and interventions such as surgery, are being severely curtailed at present, with this is likely to go on for a long period of time. The measures introduced in some countries (for example, not being able to leave home for several weeks, even for those who are not sick), will have an impact on people's mobility and it will create physical inactivity, even for short periods of time, which can increase the risk of metabolic disease.

In addition, people will have no choice but to rely on processed food which has a longer shelf life instead of fresh produce, and canned food with higher quantities of sodium, meaning we might see an increase in weight if this persists for a longer period of time.

CHAPTER 15

COVID-19 AND MENTAL HEALTH

The Coronavirus pandemic and the economic recession that came with it, due to the necessary draconian government measures to save peoples' lives, means that COVID-19 has affected many people's mental health and created new barriers for people already suffering from mental illness and substance abuse. Those who have lost loved ones or caught the virus and are worrying if they will be the next death statistic, to those who are constantly worrying about catching the virus whenever they go out on essential journeys, to those who have lost their job and are worrying how they will feed themselves and their families, to those who are feeling lonely and isolated because they live alone and have no companionship nor do they see any familiar faces, or any faces, for that matter because of the lockdown restrictions. According to kff.org, about four in ten adults in the US have reported symptoms of anxiety or depression during the pandemic. From July 2020, many adults are reporting specific negative impacts on their mental health and well-being, such as difficulty sleeping (36 percent) or eating (32 percent), increases in alcohol consumption or substance use (12 percent), and worsening chronic conditions (12 percent), due to worry and stress over the coronavirus.

Young adults in the United States have also experienced mental health issues during the pandemic. COVID-19 related problems such as the closure of universities and loss of income, due to loss of jobs or being made furloughed, have contributed to poor mental health in young people. During the pandemic, a larger share of young adults (18–24) reported symptoms of anxiety and/or depressive disorder (56 percent). Compared to all adults, young adults are more likely to report substance use (25 percent vs. 13 percent) and suicidal thoughts (26 percent vs. 11 percent). Prior to the pandemic, young adults were already at a higher risk of poor mental health and substance use disorder, though many did not receive treatment. Regarding suicidal thoughts, an 11-year-old boy died after he shot himself in his California home during a Zoom class this week, according to a new report. The pre-teen was attending the virtual class with his microphone and camera turned off when he took his own life.

Many essential workers continue to face several challenges, including the greater risk of contracting the coronavirus than other workers. Compared to nonessential workers, essential workers are more likely to report symptoms of anxiety or depressive disorder (42 percent vs. 30 percent), starting or increasing substance use (25 percent and 11 percent), and suicidal thoughts (22 percent and 8 percent) during the pandemic.

According to health.org.uk, more than two-thirds of adults in the UK (69 percent) have reported feeling somewhat, or very, worried about the effect COVID-19 is having on their life. The most common issues affecting people's wellbeing is worrying about the future (63 percent), feeling stressed or anxious (56 percent) and feeling bored (49 percent). While some degree of worry is understandably widespread, more severe mental ill health is being experienced by some groups.

Analysis of longitudinal data from the Understanding Society study found that, taking account of pre-pandemic trajectories, mental health has worsened substantially (by 8.1 percent on average) as a result of the pandemic. Groups have not been equally impacted; young adults and women who had worse mental health pre-pandemic have been hit the hardest.

The UCL COVID-19 social study of 90,000 UK adults has monitored mental health symptoms throughout the lockdown, determining that levels of anxiety and depression fell in early June as lockdown measures began to lift. But these remained highest among young people, those with lower household income, people with a diagnosed mental illness, people living with children, and people living in urban areas.

Lockdown has brought social isolation to many, particularly people living alone or those who have been shielding. People are shielding because they either have a disability or are vulnerable health-wise or due to their fragility of old age. Social isolation is an objective measure, which may or may not lead to the subjective feeling of loneliness. Perhaps, surprisingly, the proportion of people reporting that they feel lonely often or always during lockdown has been similar to pre-pandemic, around 5 percent (2.6 million) during April.

However, social isolation and national lockdowns have the potential to have detrimental effects other than loneliness and depression. There have been serious concerns about women experiencing domestic abuse. Most women who have experienced this issue have either developed mental health issues due to increased anxiety over their safety, or they have come out dead from their house in a body bag. There was a 49 percent increase in calls to the national helpline Refuge during the lockdowns.

One young person expressed their concern to the mental health charity Mind about the pandemic saying, “I’m constantly feeling helpless and frustrated, and hate the idea of anyone around me being hurt or dying. The lockdown is the biggest problem because I rely on being able to see the people I love as a coping mechanism for my anxiety and depression." This is what has affected peoples' mental health significantly, the fact that they cannot see loved ones that can help them to calm down when they start feeling stressed or when they have an episode. According to Mind, just over two thirds (68 percent) of young people said that their mental health had got worse during the lockdowns, with this rising to three quarters (74 percent) of people aged 18–24. Two thirds (65 percent) of adults and three quarters (75 percent) of young people who have experienced mental health problems said their mental health has gotten worse during the lockdowns. Over half of adults (51 percent) and young people (55 percent) who have not previously experienced mental health problems also said their mental health has gotten worse during this period. Another young person voiced their concern about the lockdown by saying, "We all thought the lockdown would be over by now, and things would be getting back to normal. But it feels like the longer this last, the more hopeless everyone is getting. I've got nothing to look forward to, so what's the point in keep going?"

Take for example the story of Carly who lives in Newcastle Upon Tyne with her 26-year-old son. For someone like her, she says lockdown, "Has been a nightmare". Carly suffers from OCD and health-related anxiety. She also has a range of physical health problems, suffering from Fibromyalgia, Myalgic Encephalopathy and Osteoarthritis of her hip, and was diagnosed with autism three years ago. "It's like being imprisoned in your own home," Carly said. Unable to get out and about without aid, she was unable to enjoy the daily outdoor exercise breaks many people took advantage of.

According to Mind, people living in social housing (a proxy for social deprivation) were more likely to have poor mental health and to have seen it get worse during the pandemic. Over half (52 percent) of people living in social housing said their mental health was poor or very poor (and 38 percent said it was not poor or very poor) and over two thirds (67 percent) said that their mental health got worse during lockdown. Similarly, over half (58 percent) of under-18s who have received free school meals said their mental health was poor or very poor, with nearly three quarters (73 percent) of this group saying that it got worse during lockdown.

Take the story of Brian, a Black male with bipolar disorder in his late thirties, who spent the pandemic alone in his one-bedroom flat in East London, Stratford, which he has been renting for nearly fifteen years. Prior to the lockdown, Brian was working in retail for a local company, however, he was since made redundant with no option of furlough. Left feeling unsure of where to turn, Brian applied for Universal Credit and was forced to contemplate spending the next few months on an income which would 'barely cover rent and food bills' while looking for a new job.

During the pandemic, people working on the frontline have experienced mental health issues like depression, substance abuse and post traumatic disorder from working during the pandemic, worried that they too might catch the virus, coupled with the fact that medical staff are banned from taking their regular annual leave, which leads to anxiety. Some doctors and nurses have to carry mobile phones or tablets to patients' bedsides and hold it up so that the patients can talk to their loved ones, which creates emotional trauma for the staff. Police officers working with the public to maintain law and order and to enforce the lockdown rules, knowing that they could be about to tackle a COVID-19 positive person who is violating the lockdown rules to the ground who might get violent with them thereby putting them at risk. Teachers who have no choice but to go to their schools to teach children of key workers who are out there working to offer priceless services to the nation. Supermarket workers that are working to ensure that the nation continues to get their food supplies during the lockdown. All these people have developed some kind of mental health problem due to the fact that they are worried that while others are in the safety of their homes with their COVID safe bubble running from the virus, they are out there risking their lives by going into the midst of the virus. Also, they are worried that they could bring the virus home to their families. Key workers like carers had to result to living temporarily at the care homes where they work and not go home on a daily basis to their family because they wanted to protect their love ones. This had a significant mental impact on these carers because they were away from their family not knowing if they were alright.

According to Mind, over half (54 percent) of parents with children under 18 said that looking after children or family members in the home made their mental health worse. They were also more likely to be negatively affected by their work situation (60 percent with children and 52 percent without) or their financial situation (53 percent and 43 percent). According to Vanessa, a 40-year-old woman from Manchester, United Kingdom, "I'm a single parent with kids in a small flat, with limited amount of time outdoors as well as the pressure of trying to maintain my job from home. It's difficult and I'm fed up." People with children experienced a lot of stress during the lockdown when working from home as they have to look after their children, supervise them with their home schooling and render all the regular parental duties, while working on their computers and dealing with clients over the phone or online.

The children's charity, Barnardo's, have voiced out concerns in their report dated 22nd June 2020, that there is a rise in the number of children requiring foster care during the pandemic. Around 52 percent of children need this vital child protection service during the pandemic. Before the pandemic, there were 11,000 children who needed this service. Now that number has rose to around 16,000 children. This rise in children needing fostering is due to the increase in mental health and domestic violence from one or both parents during the pandemic.

Across the world, even for those not directly impacted by COVID-19, the effects of social isolation and economic fallout are being felt widely. Large sections of populations around the world are experiencing increased anxiety, depression, stress, and loneliness as a result of the COVID-19 outbreak. According to the World Health Organization, the COVID-19 pandemic has disrupted, or halted, critical mental health services in 93 percent of countries worldwide, while the demand for mental health is increasing, according to a new WHO survey. The survey of 130 countries provides the first global data showing the devastating impact of COVID-19 on access to mental health services and underscores the urgent need for increased funding.

The pandemic is increasing demand for mental health services. Bereavement, isolation, loss of income and fear, are triggering mental health conditions or are exacerbating existing ones. Many people may be facing increased levels of alcohol and drug use, insomnia, and anxiety. Meanwhile, COVID-19 itself can lead to neurological and mental complications, such as delirium, agitation, and strokes. People with pre-existing mental, neurological or substance use disorders are also more vulnerable to the SARS-CoV-2 infection, and they may be at a higher risk of severe outcomes and even death.

Data from the World Health Organization shows that over 60 percent of institutions reported disruptions to mental health services for vulnerable people, including children and adolescents (72 percent), older adults (70 percent), and women requiring antenatal or postnatal services (61 percent). Sixty-seven percent of institutions saw disruptions to counselling and psychotherapy; 65 percent to critical harm reduction services; and 45 percent to opioid agonist maintenance treatment for opioid dependence. More than a third (35 percent) reported disruptions to emergency interventions, including those for people experiencing prolonged seizures; severe substance use withdrawal syndromes, and delirium, often a sign of a serious underlying medical condition. Thirty percent reported disruptions to access to medications for mental, neurological and substance use disorders. Around three-quarters reported at least partial disruptions to school and workplace mental health services (78 percent and 75 percent, respectively).

Poorer countries are feeling the impact from the pandemic the most. The COVID-19 pandemic has exposed the already large treatment gap in mental health across low to middle income countries and threatens to widen it. New demands for mental health care in these countries intersect with fragile health systems, scarce resources and workforce capacity, social unrest and violence in response to COVID-19 containment strategies, and overall scarce and inequitable access to evidence-based interventions. It can be speculated that the long-term consequences on mental health will be particularly severe in the lowest resourced and most impoverished regions of the globe, where there was virtually no access to mental health services before the pandemic. More than 10,000 individuals in Bangladesh reported a 33 percent prevalence of depression and a 5 percent prevalence of suicidal ideation.

As poverty and socioeconomic inequities are prominent in low- and middle-income countries (LMICs), and with poor coverage of adequately resourced healthcare and social safety nets, it is highly plausible to expect mental health problems in large sections of communities across these countries. For example, this pattern is already apparent in Brazil where the greatest risk of disease transmission is among the poorest communities in the country.

CHAPTER 16

NEW CORONAVIRUS UK GOVERNMENT'S ACTS

The Coronavirus Act 2020 received Royal Assent on 25th March, having been fast-tracked through parliament in just four sitting days. The Act contains **'emergency powers'** to enable public bodies to respond to the COVID-19 pandemic. The Act has three main aims: to give further powers to the government to slow the spread of the virus, to reduce the resourcing and administrative burden on public bodies, and to limit the impact of potential staffing shortages on the delivery of public services.

According to the Institute for Government, there are several acts the UK government passed under the Coronavirus Emergency Bill at the start of the pandemic:

Mitigating NHS staffing shortages

The Act enables the registration of recently retired health and social care professionals, medical students near the end of their training, and those who have recently left the profession. The Act suspends restrictions on the number of hours retired staff who return to the NHS can work. It enables volunteers in the health and social care sectors to take unpaid leave with a

UK-wide compensation fund. The Act also provides provision to facilitate emergency volunteering. This act made it possible for the government to use final year medical students to help ease the pressure on NHS staff by throwing these nearly qualified medical doctors into the midst of the virus. Retirees are re-employed into the service, and qualified doctors and nurses that were about to be deported, have their visas extended and are re-employed.

Easing pressure on NHS and local authority resources

The Act allows NHS providers to delay the assessment of a patient's need for ongoing nursing care before discharging. The Act eases, in exceptional circumstances, the requirements on local authorities to conduct a **'needs assessment'** when it appears that an adult may have needs for care and support. The Act allows for powers to detain and treat patients for mental health disorders to be implemented using the opinion of fewer medical professionals.

Reducing administrative burden on frontline staff

The Act eases the regulations relating to the registration and certification of deaths and still-births, and permissions to conduct cremations. The Act removes the requirement that any inquest into a death from coronavirus be held with a jury in England, Wales and Northern Ireland (as is required by law for other notifiable diseases).

Indemnity

The Act enables the secretary of state and ministers in devolved administrations to provide an indemnity for clinical negligence liabilities arising from NHS activities.

Management of dead bodies

A national authority or designated local authority has the power to require organisations to provide facilities, premises, vehicles, or services to manage capacity problems in the transportation, storage, and disposal of dead bodies.

Modifying requirements under the Investigatory Powers Act

Warrants under the investigatory powers act must be signed by the secretary of state and one of fifteen judicial commissioners. Because COVID-19-related sickness may result in a shortage of commissioners, the Act will allow additional judicial commissioners to be appointed on a temporary basis and the appointments process to be amended. Usually, judicial commissioners must retrospectively approve a warrant within three days of it being made. To relieve pressure on commissioners, the Act allows this period to be extended to a maximum of twelve days. Warrants are usually valid for a maximum of five days. The bill extends this period to a maximum of twelve days.

Extension of time limits for retention of fingerprints and DNA profiles

The Act allows the government to extend the period for which fingerprints and DNA profiles may be retained for up to six months if the secretary of state considers that coronavirus is having, or is likely to have, an adverse effect on the capacity of those responsible for national security decisions, and it is in the interests of national security to retain fingerprints or DNA profiles.

Suspending port operations

The Act provides powers to suspend port operations if shortages in Border Force staff mean there are insufficient resources to secure the border. Initial decisions to suspend port operations can be taken by senior Border Force Officials on behalf of the Secretary of State. Suspensions for more than twelve hours must be taken by ministers.

Powers relating to potentially infectious persons

The government has already passed secondary legislation to give public officials in England emergency powers to test, isolate and detain a person where they have reasonable grounds to think that the person is infected. The Act puts those powers on a statutory footing and extends them to authorities across the whole UK.

Someone who breaches a direction given under these powers commits an offence and is punishable by a fine.

Powers regarding public gatherings and premises

The Act gives ministers, including in the devolved administrations, the power to restrict or prohibit gatherings or events, and the power to close or restrict access to premises. The minister can only use this power if they have made an official declaration that the virus constitutes a "serious and imminent" threat to public health, and that using the powers would either help to control the transmission of the virus or would facilitate the most appropriate deployment of medical/emergency resources. Someone who breaches such a direction commits an offence, punishable by a fine.

Vaccinations in Scotland

At present, only medical practitioners or those acting under their control can administer vaccinations in Scotland. The Act allows a wider range of health professionals to do so.

Schools/childcare providers

The Act gives ministers, including the devolved administrations, the power to require the temporary closure of a school or registered childcare provider. When a minister has given such a direction, the institution must take reasonable steps to stop people attending the premises for a specified period. The minister can also make more specific directions about particular parts of the premises or particular people. Ministers have to take advice from public health officials before using these powers.

Technology in court

The Act makes several provisions for parties and witnesses in court proceedings to appear by live link, rather than in person. It also provides that, where someone is appealing a government decision to restrict their activities to the magistrates' court, they appear by live link.

Statutory sick pay

The Act enables the government to make regulations to allow certain employers to reclaim the cost of providing statutory sick pay to their employees from HMRC for COVID-19-related absences. It also makes statutory sick pay payable from day one, rather than day four, of sickness. The government hopes this will remove a disincentive to workers staying at home when they are infected.

Pensions

Rules preventing those in receipt of NHS pensions returning to work will be suspended. The government hopes this will remove a disincentive to retired health professionals returning to work.

National Insurance Contributions

The Act will temporarily reduce the requirements for changing rates of National Insurance contributions.

Protecting tenants

Residential tenancies protection from eviction: in the Commons, the government introduced new provisions extending the statutory notice period for evictions for most residential tenancies from two months to three months.

Business tenancies protection from forfeiture: in the Commons, the government introduced new provisions temporarily restricting the ability to enforce re-entry or forfeiture for non-payment of rent.

Food supply

The Act gives the government power to require food suppliers and retailers to provide information relating to food supply chains. This act helped many families that were suffering financially through the loss of job or being made furlough, to be able to have food provisions from the government and organisations like the British Red Cross, Food Banks, and much more on a regular basis.

How long measures will last for?

Most of the Act will stop having an effect two years after it is passed. Some provisions, including certain provisions relating to the emergency registration of health professionals and indemnity of health service activity, do not expire after two years.

Following the governments amendments in the Commons, MPs will now have an opportunity to express a view on the continued operation of the Act's temporary provisions every six months. Every six months, a minister must, 'as far as practicable', make arrangements for MPs to vote to keep the provisions of the Act in force. If MPs are able to vote against keeping the provisions of the Act in force, the government must make regulations to prevent provisions having an effect within 21 days. MPs will only be able to vote on the continuation of the powers if parliament is sitting. If they are not able to vote, the powers will remain in force.

CHAPTER 17

HOW COVID-19 CHANGED THE WORLD?

COVID-19 has exposed the shaky foundations in our societies, especially regarding what we take for granted in developed countries. From the intricately interwoven nature of globalisation that enables supply chains and manufacturing infrastructure, to the just-in-time deliveries to supermarkets, as well as stark contrasts between nationalised healthcare systems such as the NHS and those financed by private insurance, the virus has revealed the flaws in each sector, thereby introducing revolutions into our world. Some of these changes we are familiar with already, like using robots in the manufacturing industry to make cars and so on, but the virus will now increase this pattern to more industries around the world. Other patterns that the virus may introduce are totally new to the world and people will have to adjust themselves to get use to this new lifestyle.

With the tremendous amount of fear that has come with the Coronavirus pandemic such as social anxiety, which is a disorder that creates an overwhelming fear of social situations like meeting or speaking to people, dreading social events, avoiding eye contact and low self-esteem. People are scared of each other and of their own loved ones and are becoming reclusive and afraid to go outside of their house due to COVID-19, because no one knows if the next man or woman they come across has the virus. This fear is speeding up the latest wave of automation which already existed before the pandemic but is now spiralling out of control due to the virus, because pretty soon no one will be in offices or warehouses and construction sites because they are at home or in the hospital sick with the virus, or they are dead from it.

Robotization is becoming more common in restaurants, factories, warehouses and in other businesses, in a frenzy to reduce risk of catching the virus and allowing businesses to save money on labour costs. All of that is post-industrial. But we are also now experiencing a shift back to the Pre–Industrial Age, with large parts of the economy based in homes and vehicles. Both workers and their employers are becoming accustomed to the work-from-home movement, and much has already been said about how this jump seems permanent. What has been discussed less is the coming reverberation in cities, built up over centuries into metropolises of gigantic office and residential buildings whose valuations could change dramatically because soon, they will become useless as no one will require a physical office or shop store anymore as everything is done online from home. It is hard to imagine a repeat of the age of the plague, when the answer was that poor people from the countryside moved in. But new uses will have to emerge for lesser occupied, if not abandoned, office buildings.

With the introduction to working from home, we are asked to acquiesce to a different kind of intrusion: software that allow companies to monitor who is actually working. That is no accident. The post-virus world is likely to be ever more Orwellian. For the first time in history, governments can actively survey and respond to everyone and punish those who defy public ordinances such as health orders. Just as people have come to expect cameras recording their movements on the street since 9/11, people in the post-COVID-19 world, that is if the virus is eradicated, may see nothing unusual about more intimate measures like public monitoring of their temperature and blood pressure. Working from home means logging into a specific system where your managers can monitor your progress on assignments and inspect your living-room while you work.

During the pandemic, the virus has severely disturbed arenas with the highest overall physical proximity scores. Arenas like hospitals, care environments, personal care arenas, on-site customer service, and leisure and travel arenas. In the longer term, work arenas with higher physical proximity scores are also likely to be more unsettled, although proximity is not the only explanation. For instance, the on-site customer interaction arena includes frontline workers who interact with customers in retail stores, banks, and post offices, among other places. Work in this arena is defined by frequent interaction with strangers and requires on-site presence. Some work in this arena migrated to e-commerce and other digital transactions, a behavioural change that is likely to stick.

The leisure and travel arenas are home to customer-facing workers in hotels, restaurants, airports, and entertainment venues. Workers in this arena interact daily with crowds of new people. COVID-19 forced most leisure venues to close in 2020 and airports and airlines to operate on a severely limited basis. In the longer term, the shift to remote work and related reduction in business travel, as well as automation of some occupations, such as food service roles, may curtail labour demand in this arena. The computer-based office work arena includes offices of all sizes and administrative workspaces in hospitals, courts, and factories. Work in this arena requires only moderate physical proximity to others and a moderate number of human interactions. This is the largest arena in advanced economies, accounting for roughly one-third of employment. Nearly all potential remote work is within this arena.

The aftermath of this coronavirus pandemic will also see myriad changes, from personal adjustments to global shifts. But which of these changes will have a lasting impact and which might we never see again? To answer that we need to look at how we have already begun to adjust. Changes in our personal lives, which we all had to make during the several lockdowns might last forever. It is likely that all of us experienced the imposition of lockdown as a shock to the system, whether it made us feel lonely or listless or anxious or driven to distraction by the family constantly under each other's heels, or all of the above, all at the same time. As individuals, we have had to make changes both big and small to our everyday lives.

But while physically distanced, the internet and social media have allowed us to reach into each other's homes over the countless weeks. Social relationships for many seem not to have suffered. They have also allowed us to explore hobbies and interests we might never have had before, like the people turning to social media to solve real-life mysteries from their homes.

Crises also invariably nurture the emergence of a great common purpose, solidarity, creativity, and improvisation. Social media has opened little window into how everybody else has responded and found their own coping mechanisms. Shortages of commonplace items, or difficulties in getting out to the shops or securing a delivery slot, or perhaps just that many of us have more time on our hands these days, has unlocked an inner creativity and resourcefulness that can be shared widely online. This has manifested itself in different ways. Many of us are now taking a lot more time and consideration over cooking. Not just picking up a microwaveable dinner from the supermarket on the way back from the office, but actually cooking for ourselves, carefully choosing a recipe, chopping and stirring ingredients, grinding spices, taking delight in the process of making a meal.

We may see the adoption of temperature checks or thermal imaging cameras in the entrance foyer of larger office blocks to send-home anyone showing signs of a fever (although there are doubts over the actual effectiveness of such screening technology). And workplaces previously using hot-desking will likely need to reconsider their arrangements. Bustling offices with multiple people using the same desk space would be hotbeds for transmission. Many businesses may also need to stagger work-shifts so that offices and factories do not become overcrowded, and workers can safely maintain distancing. This is likely to cause a reduction in rush hour traffic, with commuters no longer needing to travel to and from work at the same time.

Also, there is the online trend which has picked up more so now during the pandemic than it was before. Even though supermarkets were opened throughout the lockdowns, people preferred buying online in order to limit the risk of catching the virus. School lessons are conducted online using Zoom or Microsoft Teams to teach children about Geography, Maths, English, Science and so on. People can see a medical doctor or any other health professional through a video link where they can assess you, just like they would in a face-to-face session and diagnose your health problems. Programmes like Telehealth, which provide and manage health care in which individuals manage aspects of their care with remote support from health-care professionals. Care is most digitally mediated but supported by direct communications from staff.

Behind all the suffering and disruption and economic hardship of the Coronavirus pandemic, an even larger global crisis is lurking in the midst: climate change. Could our experiences with the international lockdowns help the environmental cause, or would we just return to "business as usual" as quickly as possible? Many city-dwellers have noticed an improvement in their urban environments with cleaner-smelling air, calmer, safer roads, and bolder wildlife, which offers a glimpse of what a greener world might be like to live in. In some way, lockdown has helped a bit by reducing air pollution which in itself is a global pandemic, as it is detrimental to our health, as everyone is locked inside their homes afraid to venture outside not to catch the virus or incur a fine for breaking lockdown rules, which means there are fewer motor vehicles on the streets. This will help in cleaning the air we breathe and make our health better.

Young, working people are concerned that due to the frequent halt in society because of the lockdowns and social restrictions, along with the limitations caused by Long COVID, their financial circumstances have worsened. The amount they could have saved for the future for things like paying towards their dream house, buying a car, or paying for their kids' University study, going on a dream holiday, and much more, have been affected by the pandemic and its aftermaths. Long COVID has affected many people which means that these once hard-working people can no longer work the way they used to before. With symptoms like extreme tiredness, shortness of breath, chest pain and tightness, problems with memory and concentration, which is known as brain fog, insomnia, dizziness, heart palpitation and much more, just how can these people work or even manage to get back to their normal work? It is certain that the unemployment rate will drastically increase in the UK and the rest of the world, with a 98 percent increase in Universal Credit claims in January 2021 (6 million new claims), according to Gov.uk. Among those who say their financial situation has gotten worse during the pandemic, 44 percent in the United States think it will take them three years or more to get back to where they were a year ago, including about one-in-ten who do not think their finances will ever recover. COVID-19 has also caused housing demand to be high across the world since it started.

Also, the National Health Service, along with similar health services across the world, have been severely affected by the pandemic, causing many people with other serious health conditions to not come forward for treatments. People who were due to have surgery, are now in hiding due to the fact that they are afraid to visit the hospitals as they do not want to catch COVID. This is causing a huge problem for the NHS as the backlog continues to pile up to the level where it might become unmanageable. This is also creating a huge amount of stress for NHS staff as they have to deal with ten times the workload that they are used to, along with dealing with the virus and how to keep themselves safe while on duty. This backlog is believed to last a while after COVID has subsided, which will significantly affect the health system.

CHAPTER 18

LEARNING FROM THE MISTAKES: W.H.O AND UK GOVERNMENT'S ERRORS

Boris Johnson said that his government did everything they could in the fight against the Coronavirus. The Prime Minister said during the daily Coronavirus press briefing that he was, "Deeply sorry for every life that has been lost." He also said that "It's hard to compute the sorrow contained in the statistics. The years of life lost, the family gatherings not attended and for so many relatives, the missed chance to say goodbye."

Prime Minister Boris Johnson has announced to Parliament that he is launching a public inquiry to scrutinise the way the government handled the entire Coronavirus pandemic, using a wealth of evidence that will come from bereaved families, survivors, scientists and much more. The inquiry is said to take place in spring 2022. According to the Guardian newspaper, Mr. Johnson said it was "Absolutely vital that we should learn the lessons of tackling COVID," promising a chair would be appointed and terms of reference confirmed after consultation with the devolved administrations in Scotland, Wales, and Northern Ireland. He admitted bereaved families might be anxious for the inquiry to start earlier than spring 2022 but said it would be wrong to weigh down scientific advisers and take up "Huge amounts of officials' time" when they may still be in the middle of the pandemic, if cases rise again in the winter.

However, there were several mistakes that the Prime Minister and his Conservative government made that caused the virus to spiral out of control in the country. For instance, he was adamant that there would be no national lockdown and what is paramount, is to get herd immunity of at least a hundred thousand people so that scientists can study the virus thoroughly through those who have become immune to the virus. This means putting one hundred thousand Brits' lives in danger of catching the disease and dying from it instead of saving their lives.

Also, the government, after wasting precious time, ordered its entire population into several national lockdowns in order to stem the flow of the virus that lasted for months, without any clear indication as to when those lockdowns will end, and yet they did not provide any financial support for families and individuals who for most, were already suffering financially and reliant solely on Universal Credit. Even though the Chancellor poured billions of pounds into the economy to help fight the COVID-19 pandemic, Mr. Sunak never allocated money for the people so that they can do their essential shopping for food and other items for their home and family before the lockdown began, so that they did not have to frequent the supermarkets unnecessarily. These are the taxpayers that have built up this economy when they were working and now, instead of their government rewarding them and helping them through the pandemic, they were left in the dark. Other countries like the United States gave each citizen $1,400 stimulus cheques to help them through the pandemic. According to Chronicle Live, an exercise codenamed 'Cygnus' in 2016, found the UK's preparedness was 'not sufficient' for a major flu outbreak. But instead of readying our COVID defences in January and February 2020, the PM celebrated Brexit, spent time at Chequers and sorted his matrimonial arrangements. By the end of January, the virus had spread from China to six countries and the World Health Organization warned of a public health emergency of international concern. Boris Johnson missed five important meetings of the Government's COBR emergency committee. On 18th February 2020, his divorce was finalised, and on 29th February 2020, Mr Johnson and Carrie Symonds announced they were engaged and having a baby in the early summer.

Government advisors also warned people should stop shaking hands to slow the spread of Coronavirus on the same day Boris Johnson boasted he had been shaking hands with everybody at a hospital. The Independent Scientific Pandemic Influenza Group on Behaviours (SPI-B), a subcommittee of the SAGE scientific advisory committee, issued its advice on 3rd March 2020. It said, "There was agreement that the government should advise against greetings such as shaking hands and hugging, given existing evidence about the importance of hand hygiene. Promoting a replacement greeting or encouraging others to politely decline a proffered hand-shake may have benefit." But on the same day, Mr. Johnson told a No 10 press conference, "I was at a hospital the other night where I think a few there were actually Coronavirus patients. And I shook hands with everybody. You'll be pleased to know, and I continue to shake hands."

Widespread Coronavirus testing was axed on 12th March 2020, as ministers decided to focus on testing people in hospitals and care homes. It was only reintroduced months later once capacity was high enough with care home residents not being routinely tested until mid-April 2020. Scientists have since said it would have been better to get widespread testing going sooner as infected people went untraced.

At the time community testing was dropped, Jenny Harries, Deputy Chief Medical Officer for England, claimed it was "Not an appropriate mechanism as we go forward." But she too, later accepted the nation would have taken a different path if more tests were available.

Mr Johnson overruled Government scientists in September who pressed for national lockdown measures such as stopping all household mixing and closing all pubs. SAGE called for an immediate introduction of national interventions, saying failure to take such measures could result in "A very large epidemic with catastrophic consequences." Top of this list of five interventions was a two- to three-week circuit breaker lockdown. Mr Johnson ignored those calls and a close ally told journalists, "Keir Starmer, who urged a circuit-break, is a shameless opportunist playing political games in the middle of a global pandemic." He eventually called a second national lockdown in November, when cases had risen much higher.

COVID-19 ripped through care homes at the start of the pandemic, as elderly people proved extremely vulnerable to the virus. Almost a third of Coronavirus deaths to date have been in care homes. The elderly and infirm are of course more at risk from the virus, but it emerged that many elderly people were discharged from hospitals into care homes in March to free up beds without being tested. This decision has been blamed for helping the virus to spread. Guidance to care homes in February 2020 said that "This guidance is intended for the current position in the UK where there is currently no transmission of COVID-19 in the community. It is therefore very unlikely that anyone receiving care in a care home, or the community will become infected." The guidance said there was no need for staff to wear masks. It was only withdrawn on 13th March 2020. The Chief Medical Officer Chris Whitty has since admitted a major failing was not realising how far the virus could be spread asymptomatically.

The Government dragged its feet for months over measures to tighten up the UK's borders which Priti Patel claimed she had privately pushed for. People were allowed to flock in and out of Britain and other countries as they wish after the end of the first lockdown, then rising case numbers in countries like Spain led to sudden, chaotic introductions of quarantine restrictions. The Government is finally considering making people quarantine in hotels. Other countries have had the policy for months.

Even at the end of February 2020, the UK Government believed it was in a strong position to help others by sending more than 600,000 items of PPE to China to help tackle the Wuhan outbreak. But in the early months of the pandemic in the UK, care and health staff complained they were running out of PPE and had to improvise using bin bags for a safety apparel. Some items that the government hastily ordered through lucrative private sector contracts turned out to be unusable or not to meet Government specifications. They included gowns which were flown in from Turkey on an RAF cargo plane in a chaotic scene and amid several delays.

Schools too had their share of government errors. Ministers insisted that schools are safe for the students and teachers who use them, and they are vital for stopping other harms such as a lost generation of learning. Even now there is no firm consensus on how much they might spread the virus in the wider community, which is the reason they can be a problem. However, they have been embroiled in chaos and chopping and changing rules. At first, they had strict 15-pupil bubbles, but these were then extended to whole year groups, essentially making them pointless as a containment strategy.

England's primary schools then were allowed to return for one day in January 2021 before then being locked down again to all but vulnerable and key workers' kids. What the government forgot is that children in schools come from different households, which breaks the government's rule of 'not mixing with different households.' It makes no sense that the government would say that "You are not allowed to see and visit loved ones," yet they allowed schools to remain open where pupils got to mix with others who might end up sneezing or coughing out COVID particles. Some teachers have gone away for holiday and have contracted the virus and are afraid to inform the school for fear that they could be asked to isolate for two weeks or even get sacked, which would impact their finances and teaching career completely. Yet, these were the same teachers that were teaching the students. Also, the student could be asymptomatic to COVID and spread it to the teachers and other students.

To make matters even worse, the Prime Minister's former chief adviser, Dominic Cummings delivered some harsh blows on Wednesday 26th May 2021, critiquing Mr. Johnson and his government's handling of the pandemic, saying that the PM is unfit to run the country and that the Health Secretary, Matt Hancock, should have been sacked many times for "Lying to everybody on multiple occasions in meeting after meeting in the cabinet room and publicly." Mr. Cummings told the cross-party Health and Science and Technology Committees, chaired by the former Conservative cabinet ministers Greg Clark and Jeremy Hunt, that the Prime Minister took the outbreak of COVID-19 as a joke, saying that it is just another "Scare story being circulated by the media and that COVID is just like Swine Flu," and he will get Chris Witty, who is England's Chief Medical Officer, to inject him with the virus on national TV so that people can see it has no detriment on their lives.

Mr. Cummings also told the committee that Helen McNamara, the Deputy Cabinet Secretary, informed him and other government advisers there was "No plan" to deal with the severity of the threat. The threat being COVID-19, which was already ravaging Wuhan and other countries. Cummings also said that Mark Sedwill, then head of the Civil Service, advised the prime minister to appear on television and tell the public that COVID "Is like chickenpox" and they needed to have "Chickenpox parties" to spread the virus in order to get the herd immunity, which was suggested by the PM. He then went on to tell lawmakers that the Prime Minister said that he would rather see bodies pile up than put the country into a third national lockdown. All this evidence came months before the public inquiry will begin in 2022, which has been proposed by Mr. Johnson.

The World Health Organization (WHO) also made its own blunders at the start of the pandemic. According to First Post, the Independent Panel for Pandemic Preparedness and Response (IPPPR) said a series of bad decisions meant that COVID-19 went on to kill at least 3.3 million people so far and devastated the global economy. Institutions "Failed to protect people" and science-denying leaders eroded public trust in health interventions. Early responses to the outbreak detected in Wuhan, China in December 2019 "lacked urgency", with February 2020 a costly "lost month" as countries failed to heed the alarm, said the panel. The panel, which was jointly chaired by former New Zealand's Prime Minister Helen Clark and former Liberian President Ellen Johnson Sirleaf, did not spare the World Health Organization, which requested it to conduct the inquiry, and in their report which is titled: COVID-19: Make it the Last Pandemic, saying that the World Health Organization "Could have declared the situation a Public Health Emergency of

International Concern (PHEIC), its highest level of alarm on 22nd January 2020. Instead, it waited eight more days before doing so. Poor strategic choices, unwillingness to tackle inequalities and an uncoordinated system created a toxic cocktail which allowed the pandemic to turn into a catastrophic human crisis." The threat of a pandemic had been overlooked and countries were woefully unprepared to deal with one.

CHAPTER 19

IS IT FAIR?

Since the start of the pandemic, people's liberties across the world have been taken away from them with all the Coronavirus restrictions that have been imposed by various governments around the world. Restrictions like social distancing, stay-at-home, wearing face masks in public, not allowed to visit friends and family, limited guests at funerals and weddings, which at one point were not even allowed to take place. However, despite all these restrictions, some people or events seemed to be exempt from the general rules.

While people are forbidden to hug their loved ones at funerals and weddings, and guests are limited to a certain amount, which means that all the loved ones that should have attended the funeral to pay their last respects or attend the wedding or graduation ceremony of a loved one to celebrate together with them, cannot do so. Yet, football fans across Europe and South America can congregate in their thousands to support their countries in the Euro 2020, which was postponed due to the pandemic, and Copa America 2021, with hardly any social distancing, and they can hug other fans when their team scores. Fans can travel across borders to watch their team play. Scotland fans were able to flock into Wembley stadium to watch Scotland play against England.

More than 60,000 fans will be allowed to attend the semi-finals and final of Euro 2020 at Wembley Stadium, the government have confirmed. After attempts from other nations, including Italy and Hungary, to move the final three games from London, it was confirmed that capacity would be increased to 75 percent at the national stadium for the final three games of the tournament. VIPs can travel from one country to another to watch these matches without quarantining whilst the regular traveller is being forced (not asked) to quarantine for weeks and pay thousands of pounds to stay in airport hotels, designated by the government. By the end of their quarantine, the money they would have paid is more than enough to go on several luxurious holidays. Is this fair on the regular travellers? By no means. Rules are rules and they should apply to everyone regardless of their status and wealth in society.

Also, Wimbledon championship tennis 2021 is allowed to take place, while its French counterpart, Roland Garros has delayed its opening by two more weeks. According to Wimbledon's website, "We welcome the announcements from the Prime Minister and Culture Secretary that a number of events, including The Championships 2021, will be able to take place with higher spectator capacities than the current Step 3 guidance as part of the next phase of the government's Event Research Programme." Wimbledon has also announced that the government has permitted them to open the tournament which will commence on Monday 28th June 2021, with a 50 percent capacity (7,500 people) across the grounds, building up to full capacity crowds of 15,000 fans on Centre Court for the finals weekend. Tennis lovers, however, will not be allowed in without a negative COVID certificate according to the organisers.

Along with the Euro football and Wimbledon tennis, the World Championship Snooker also took place in April 2021 in Sheffield at the Crucible Theatre. There was a live audience at the theatre during the midst of the pandemic, and all those that attended must take a COVID-19 test and present a negative test result to enter the grand arena. This Crucible tournament along with the FA and Euro football, Wimbledon tennis, and even music events, and the Brits Awards, are all part of a government's pilot scheme to find a way out of lockdown. There were only 28 confirmed cases detected across the nine government's pilot events of a capacity of 58,000 fans, and yet the government has literally shut down the music industry while the sport industry is allowed to continue without providing the data from their pilot scheme. Is this fair on the music industry and other industries that have made sacrifices to stem the flow of the virus, and yet got a slap in the face from the government? Is this also fair on the rest of the population who have abided by the government's rules since COVID-19 started, and these people have made all the sacrifices in order to combat the virus by not seeing their loved ones for months and much more, and yet the government just opened all these events in the name of researching their way out of lockdown, packed full of fans, not following a single COVID rule?

The Tokyo Olympic Games are about to kick start in July 2021 and the event is said to have a record of 33 competitions and 339 events held across 42 competition venues, all happening during the pandemic which could spread the virus further and make a deadly situation even worse as there will be spectators and athletes coming from different parts of the world, along with the international press and much more, all of whom could be carrying the virus into the Olympic games and distributing it amongst others. Again, is it fair on the many people in Japan and around the world who have made all the sacrifices to end COVID-19 and yet these governments value building their economy more than protecting peoples' lives?

Take the business sector for example. Many companies, either small, medium, or large, have also been treated unfairly by the government. After abiding to every government's rule on COVID-19, such as closing down for months during the lockdowns, restricting the amount of customers that can be in their premises at one time when they opened up again, to marking up 2 metre signs on the floor to maintain social distancing, employing COVID marshals to enforce the rules, and providing hand sanitizer for everyone that attends their business premises, along with the provision of PPE for those that work there, which has costed these businesses lots of money, which has obviously caused their profits to decline drastically; the government gave UEFA the green light without any restrictions whatsoever. Yet, businesses are in limbo fighting to stay afloat and survive the year.

On 25th June 2021, the UK's Health Secretary, Matt Hancock came under fire after he was caught having an affair with one of his married advisers and close friend Gina Coladangelo, at work. Mr. Hancock was seen hugging and kissing her during the time when hugging was banned, and social distancing of 2-metres was being enforced by the government. Mr. Hancock did not only break the ministerial code by having an affair at work, but he broke the Coronavirus rules which he himself advocated for. Is it then fair on the British people who were told not to hug and visit their loved ones? Grandparents were forbidden from hugging their grandchildren and in some cases, they could not even see the grand kids. Fathers were not allowed to hug and carry their newly born babies, couples that live in separate bubbles could not hug and kiss each other or even see each other, and loved ones were not allowed to hug and console their family and friends during a funeral service, yet Matt Hancock was allowed to break the very rule he set. Is he above the law? By no means. On 26th June 2021, Mr. Hancock finally resigned from his job as Health Secretary instead of being sacked by the Prime Minister, and a couple of days later, he was replaced by Sajid Javid. His mistress also left the department for health on the same day.

Is it fair on the parents of Ollie Bibby or even on Ollie who died of Leukaemia at age 27, the day before Matt Hancock was caught breaking the very COVID-19 rules that he helped to put in place? Mrs Bibby said that she was “Livid he broke social distancing rules”, as she was prevented from being with her son before he died. While he was dying at University College London Hospital, her son felt like he was “In prison”, she said, and begged to see his family, but they were “Treated like criminals” and barely allowed in due to COVID-19 restrictions. Many families like Ollie’s were prevented from seeing their loved ones in hospitals while they were sick, maybe not even from the virus, and sadly for some, they were prevented from saying goodbyes to their loved ones before they died just like in the case of the Bibby’s. Yet, the man who helped to put all these restrictions in place was breaking them at will. Is it fair?

Hundreds of pilots, cabin crew and travel agents are protesting in England, Scotland, and Northern Ireland as they accuse their respective governments of failing to restart travel by “Undermining the COVID-19 traffic light system.” The industry body ABTA, which is leading the ‘Travel Day of Action’ on Wednesday 23rd June 2021, argues that there are no major tourist destinations on the quarantine-free green list, and the government has urged people to avoid holidays to countries on the amber list, leaving the industry on its knees. Hundreds of thousands of jobs have been lost and the industry leaders and unions are now fearing that the travel industry will crumble under the ongoing Coronavirus restrictions if something is not done. Is it fair on the travel industry for them to ground all their flights during the lockdowns, and when the lockdowns are over, only few flights are running because of the restrictions placed on certain destinations, and yet all these sporting events, full of thousands of spectators are taking place? Does it not occur to the government that all these countries that they are putting on their red COVID hot spot list out of precaution because of new variants or the increase in cases they may have, that there is a possibility for one of these sport fans to have a false negative COVID-19 result and spread the virus to other fans in the venue or on their way to or from the sporting venue?

CHAPTER 20

COVID-19 CONSPIRACIES

From the idea that drinking bleach can kill the novel Coronavirus to a theory that the virus was created in a lab as a bioweapon, to the claim that it stemmed from the introduction of 5G, the COVID-19 pandemic has generated a wave of misinformation. Indeed, one study, published on 10th August 2020, in the American Journal of Tropical Medicine and Hygiene, found that the pandemic has hatched more than 2,000 rumours, conspiracy theories and reports of discrimination. Some even believe that the pandemic was caused intentionally by Chinese scientists to be used as a bioweapon. According to Scientific America, a document written by Chinese scientists and health officials in 2015 before the pandemic erupted, states that SARS Coronaviruses were a "New era of genetic weapons" that could be artificially manipulated into an emerging human disease virus, then weaponised and unleashed.

5G causing COVID-19.

Since the start of the pandemic, people have been saying that COVID was created by the introduction of 5G. This conspiracy theory should be easy to debunk as it is biologically impossible for viruses to spread using the electromagnetic spectrum found in 5G mast towers. 5G is made up of waves and photons while the virus is composed of biological particles and proteins and nucleic acids. Also, because 5G was rapidly rolled out when the pandemic was at its infancy and deadliest point, conspirators say that is why the virus is so brutal. This conspiracy is being promoted by anti-vaxxers who have long been spreading fears about electromagnetic radiation. It is worth repeating what the World Health Organization say, "That viruses cannot travel on mobile networks, and that COVID-19 is spreading rapidly in many countries that do not have 5G networks." Countries like Bolivia, Iran, Sweden, and many more.

Bill Gates and his depopulation plan

Microsoft co-founder and billionaire philanthropist Bill Gates is one of the leading public figures in the fight against the Coronavirus pandemic but has also become the target of several conspiracy theories after gently criticizing the defunding of the World Health Organization and saying in a 2015 Ted talk that "If anything kills over ten million people over the next few decades, it is likely to be a highly infectious virus rather than war." This is what conspirators use to bolster their claims he had foreknowledge of the COVID-19 pandemic, or he even purposely caused it. Others also believe Gates is at the forefront of efforts to depopulate the world after he said that "The world today has 6.8 billion people. That's headed up to about 9 billion. Now, if we do a really great job on new vaccines, health care, reproductive health services, we could lower that by, perhaps, 10 or 15 percent." Gates said this in a 2010 Ted talk when he was talking about carbon emission and how reducing the world's population growth can help reduce the emission. Several others accuse him of making vaccines a requisite or even trying to implant microchips into people. A recent version of this conspiracy theory, particularly beloved by anti-vaccination activists, is the idea that COVID is part of a dastardly Gates-led plot to vaccinate the world's population. There is some truth in this, of course: vaccinating much of the world's population may well be the only way to avoid an eventual death toll in the tens of millions. But anti-vaxxers do not believe vaccines work. Instead, some have spread the myth that Gates wants to use a vaccination program to implant digital microchips into people that will somehow track and control people's lives.

The virus escaped from a Chinese lab

This rumour is the most plausible among the other conspiracies. It is true that the original epicentre of the pandemic is the Chinese city of Wuhan, which also hosts a virology institute where researchers have been studying bat Coronaviruses for a long time. One of these researchers, Shi Zhengli, a prominent virologist, who spent years collecting bat dung samples in caves and was a lead expert on the earlier SARS outbreak, was sufficiently concerned about the prospect that she spent days frantically checking lab records to see if anything had gone wrong. She admits breathing a "Sigh of relief" when genetic sequencing showed that the new SARS-CoV-2 Coronavirus did not match any of the viruses sampled and studied in the Wuhan Institute of Virology by her team.

However, the sheer coincidence of China's lead institute studying bat coronaviruses being in the same city as the origin of the COVID outbreak has proven too juicy for conspiracists to resist. The idea was seeded originally via a slick hour-long documentary produced by the Epoch Times, an English-language news outlet based in the United States, with links to the Falun Gong religious cult, that has long been persecuted by the Chinese Communist Party (CCP). The Epoch Times insists on calling COVID "The CCP virus" in all its coverage. The theory has now tipped into the mainstream, being reported in the Washington Post, the Times (UK) and many other outlets.

COVID was created as a biological weapon

It is also believed that COVID-19 did not just escape the Chinese lab, but it was intentionally created by Chinese scientists as a biowarfare weapon. Twenty-three percent of people believe it was developed intentionally, with only 6 percent believing it was an accident. According to Business Standard, there is a paper titled: The Unnatural Origin of SARS and New Species of Man-Made Viruses as Genetic Bioweapons, suggesting that Chinese military scientists were discussing the weaponization of SARS coronaviruses five years before the COVID-19 pandemic began.

The US military imported COVID into China

The Chinese government responded to the anti-China theories with a conspiracy theory of their own that seeks to turn the blame back around onto the United States. This idea was spread initially by Chinese foreign ministry spokesman Zhao Lijian, who tweeted saying, "It's possible that the US military brought the virus to Wuhan." These are rumours that are being spread all over China saying that the US army carried the virus into their country during their participation in the 2019 Military World Games in Wuhan. The Chinese even wanted to name COVID the USA virus to make their point.

GMOs are somehow to blame

Genetically Modified Organism crops have been a target of conspiracy theorists for years. First, what is a genetically modified organism? According to National Geographic, a genetically modified organism "Contains DNA that has been altered using genetic engineering. Genetically modified animals are mainly used for research purposes, while genetically modified plants are common in today's food supply." Hence, it was hardly a surprise to see it being blamed in the early stages of the COVID pandemic. In early March, Italian attorney Francesco Billota penned a bizarre article for Il Manifesto, falsely claiming that genetically modified crops cause genetic pollution that allows viruses to proliferate due to the resulting environmental imbalance. Anti-GMO activists have also tried to blame modern agriculture, which is strange, since the known path of the virus into the human population, as with Ebola, HIV, and many others, was through the very ancient practice of people capturing and killing wildlife.

On the contrary, it is possible that GMOs will be used to develop any future COVID-19 vaccine. If any of the ongoing 70 vaccine projects work (which is a big if), that would be pretty much the only guaranteed way the world can get out of the COVID mess. Vaccines could be based on either GM attenuated viruses or use antigens produced in GM insect cell lines or plants.

COVID-19 is a plandemic!

Most people around the world think that COVID-19 is not real, and it is a plandemic, fabricated by the global elites to take away our freedoms by spreading fear into mankind. Early weaker versions of this theory were prevalent on the political right in the notion that the novel coronavirus would be "No worse than Flu" and later versions are now influencing anti-lockdown protests across several states in the US, and around the world. Because believers increasingly refuse to observe social distancing measures, they could directly help to spread the epidemic further in their localities and increase the death rate.

The pandemic is being manipulated by a secret group

Some believe that a 'secret group' of America's elite is plotting to undermine the president, and that Doctor Anthony Fauci, the face of the US Coronavirus pandemic response is a secret member. Fauci's expression of disbelief when the secret group was mentioned during a press briefing supposedly gave the game away.

COVID is a plot by Big Pharma

Many conspiracy theory promoters are in reality clever actors trying to sell quack products. Alex Jones, between rants about hoaxes and the New World Order, urges viewers to buy expensive miracle pills that he claims can cure all known diseases. Dr. Mercola, a quack anti-vax and anti-GMO medic who has been banned from Google due to peddling misinformation, claims that vitamins (and numerous other products he sells) can cure or prevent COVID. Natural News, another conspiracist site, sells all manner of pills, potions, and prepper gear. These conspiracists depend for their market, on getting people to believe that evidence-based (i.e., conventional) medicine does not work and is a plot by big pharmaceutical companies to make us ill. Big Pharma conspiracies are a staple of anti-vaccination narratives, so it is hardly surprising that they have transmuted into the age of the Coronavirus.

COVID death rates are being exaggerated and hospitals are empty

Another far-right meme is the idea that COVID death rates are being exaggerated in order to instil fear into the public and therefore, there is no reason to observe lockdown regulations or other social distancing measures. Popular conspirators, who uphold this belief is Dr. Annie Bukacek, who gave the warning that COVID-19 death certificates are being manipulated by officials to ramp up beliefs of the virus. Bukacek appears in a white lab coat and with a stethoscope around her neck, making her look like an authoritative medical source. Her insistence that COVID death rates are inflated has, of course, no basis in fact. On the other hand, it is possible that the current death toll is being under-reported.

Also, you have conspirators who deliberately went into several hospitals and filmed the A&E waiting areas and corridors when they are less busy and posted this mobile phone footage on social media, claiming that the government and the News media are all telling lies about hospitals reaching full capacity and the case numbers going up. They say this because they say the hospital corridors and waiting areas are empty or not as busy as it should be if case numbers were really that high. Hospitals and the government had to respond to them to set their misinformation straight. King's College NHS Trust say the "Parts of the hospital dealing with the pandemic are extremely busy. This corridor is not representative of what is happening in wards across the hospital, which have seen a more than ten-fold increase in COVID admissions in the space of a month." These videos were filmed by individuals walking through quiet hospital corridors and are being posted on Facebook groups and on Twitter feeds of Coronavirus sceptics and anti-lockdown activists.

What do you believe? Do you think that there is a possibility for any of these conspiracies to be true? Or are they just made-up lies to put fear and doubt into people's minds about COVID and the vaccines?

EPILOGUE-WILL COVID EVER BE ERADICATED?

Coronavirus restrictions will be eased further on Monday 17th May 2021 in England, Wales and most of Scotland. The relaxation phase of the pandemic comes despite the vicious spread of the Indian variant throughout the country, along with a dire threat that is coming from Vietnam about a newly discovered variant which combines both the deadly, highly transmissible Indian variant and the UK's variant. Despite these threats of new variants arising in different corners of the world, along with the current Indian variant, people are now being allowed to visit family and friends indoor, and they can choose to hug each other if they wish. Indoor dining and all other indoor activities like watching a movie at the cinema, drinking at a pub, and so on, can resume. People are also allowed to travel overseas to the government green list countries as long as you abide by all safety measures like wearing mask and taking a COVID test. Also, football fans have also been allowed back into stadiums to watch and support their favourite teams, but only those fans who have taken the vaccine are allowed, and they must socially distance. Up to 30 people can attend weddings, receptions wakes and other life events. The number of people who can attend a funeral is determined by how many people the venue can accommodate with social distancing. Care home residents can have up to five named visitors (two at a time), provided visitors test negative.

Just when all these restrictions are being eased, the Indian variant grows worse and worse. It was reported on the same day, 17th May 2021, that there are now well over 2,000 confirmed cases of this deadly variant scattered across eighty-six boroughs, with Bolton and Bedfordshire being the epicentre for now. It seems like another lockdown is looming on the world if care is not taken. SAGE scientists issued a warning to the UK government on the 31st May 2021 that they should delay the easing of the lockdown, and instead of opening on the 21st June 2021, scientists are advocating for an additional couple of weeks of restrictions to help combat the Indian variant. On the same day, the Twickenham rugby stadium is being used to inoculate a thousand people a day. This is the largest vaccination centre in England by far.

So, will this virus ever go away so that people can get back to their normal lives? Or do we have to learn to live with this deadly virus like we have learnt to live with the Flu virus, despite having a Flu jab at our disposal. We have also learnt to live with the HIV virus despite having anti-retroviral drugs available. According to The Week, the World Health Organization's Executive Director, Michael Ryan, noted in a virtual conference that "HIV has not gone away," but that effective treatments have been developed to allow people to live with the Aids-causing virus. Contracting HIV was once a death sentence, but people with the infection can now "Live a near-normal life" thanks to antiretroviral therapy (ART) treatments that reduce the amount of the virus in the blood to undetectable levels, says the NHS.

Malaria, which is amongst some of the most dangerous viruses known to man, with an estimated 229 million confirmed cases and 409,000 deaths in 2019 worldwide, according to World Health Organization, has been in the world since the time of ancient Egypt in 1550 B.C., when similar symptoms were first reported. With all the modern scientific developments in the 21st century, the world has not been able to eradicate this virus, instead, they are only able to slow its pace down. Partial immunity in humans is developed over years of exposure, but young children remain vulnerable to this lethal disease. In 2018, under-fives accounted for 67 percent (272,000) of all malaria deaths worldwide.

It is said that there are well over a thousand variants of the Coronavirus disease waiting to emerge into the world, with each new one being more deadly than the last. Take the Lambda variant for instance, which is getting out of control in the United States and South America. It is said to be more futile, more transmissible, and more evasive than the Delta variant, and if it spreads globally, it will claim more lives than its predecessor. How can this virus, if possible, ever be eradicated? The numerous authorised COVID-19 vaccines do not seem to help with eradicating the virus either. It just helps to reduce the severity of the damage the virus can inflict on the human immune system, but with the possibility of all these variants evolving and spreading worldwide, will these vaccines even be effective in the long run, or do scientists have to keep developing new vaccines for each variant that emerges? Scientists have issued another dire warning on the 31st May 2021 saying, "UK is seeing early signs of a third wave." This is the very same warning which was issued several times before the two last waves occurred and nobody listened, which led to the large death counts.

Just how long will this go on for? How many more lockdowns will there be in the UK and the world in general? How many more people will have to sadly die or become infected? And how many more restrictions can governments impose on their citizens, and how much more billions need to be spent on scientific research and other essential services before the Coronavirus can be contained? How many more red hearts will be drawn on the National COVID Memorial Wall to represent each life that was claimed by the virus in the UK before the pandemic ends?

From the looks of things, COVID-19 is here to stay, and generations yet unborn will also suffer the same fate that we are suffering today because of the virus. The on and off lockdowns, shops (with exception to essential shops), restaurants, and all indoor venues closing, keeping two metres apart, wearing a face mask in public, checking everyone's temperature before they enter a public building and much more, will become their fate too if COVID is not eradicated. All these COVID restrictions could become the new norm of life in the 21st century and beyond. Diseases like Malaria, HIV, Flu, Ebola and much more, have stuck around for decades, affecting each new generation that arrives, and the Coronavirus is set to follow suit. The vaccines are our only hope out of this pandemic and for life to return to normal. Variants that can foil the human immune system and avoid being detected by the vaccines, which in essence will render the vaccines useless, will be an ongoing threat to our freedom and human rights.

The table below shows the latest government data for each affected country for the pandemic as of 24th August 2021. It shows the total cases, new cases, total deaths, new deaths, and total recovered.

#	Country, Other	Total Cases	New Cases	Total Deaths	New Deaths	Total Recovered	New Recovered
	World	213,273,985	+511,053	4,453,036	+7,564	190,809,781	+570,821
1	USA	38,813,582	+110,776	646,667	+406	30,568,819	+73,978
2	India	32,460,328	+11,359	435,050	+266	31,690,586	+17,483
3	Brazil	20,583,994	+13,103	574,944	+370	19,479,947	+31,131
4	Russia	6,766,541	+19,454	176,820	+776	6,034,867	+15,401
5	France	6,624,777	+5,166	113,419	+108	*6,065,321*	+26,785
6	UK	6,524,581	+31,914	131,680	+40	5,098,152	+41,581
7	Turkey	6,234,520	+18,857	54,765	+232	5,724,382	+15,833
8	Argentina	5,139,966	+6,135	110,609	+257	4,822,420	+8,400
9	Colombia	4,892,235	+2,698	124,315	+99	4,721,710	+3,210
10	Spain	4,794,352	+7,967	83,337	+67	4,165,921	+29,468
11	Iran	4,715,771	+38,657	102,648	+610	3,961,024	+28,552
12	Italy	4,488,779	+4,168	128,795	+44	4,224,429	+3,505
13	Indonesia	3,989,060	+9,604	127,214	+842	3,571,082	+24,758
14	Germany	3,881,579	+5,555	92,497	+19	3,702,100	+3,500
15	Mexico	3,225,073	+7,658	253,155	+228	2,553,626	+15,619
16	Poland	2,886,805	+107	75,316		2,656,317	+174
17	South Africa	2,698,605	+7,632	79,584	+163	2,455,998	+11,589
18	Ukraine	2,275,171	+610	53,474	+17	2,201,433	+434
19	Peru	2,142,153		197,879		N/A	N/A
20	Netherlands	1,921,159	+2,390	17,956	+4	*1,820,985*	+5,264
21	Philippines	1,857,646	+18,332	31,961	+151	1,695,335	+13,794
22	Iraq	1,832,240	+7,151	20,262	+78	1,668,310	+9,142
23	Czechia	1,677,619	+107	30,385		1,645,507	+3
24	Chile	1,634,394	+578	36,688	+38	1,590,973	+721
25	Malaysia	1,572,765	+17,672	14,342	+174	1,297,723	+19,053
26	Canada	1,473,624	+4,811	26,814	+22	1,422,576	+5,313
27	Bangladesh	1,467,715	+5,717	25,399	+117	1,372,856	+8,982
28	Japan	1,300,353	+22,285	15,631	+35	1,066,309	+14,088

Active Cases	Serious, Critical	Tot Cases/ 1M pop	Deaths/1M pop	Total Tests	Tests/1M pop	Population
18,011,168	112,251	27,361	571.3			
7,598,096	23,660	116,480	1,941	569,086,569	1,707,839	333,220,183
334,692	8,944	23,261	312	507,551,399	363,707	1,395,494,393
529,103	8,318	96,059	2,683	56,580,445	264,043	214,285,295
554,854	2,300	46,344	1,211	175,500,000	1,202,005	146,006,013
446,037	2,128	101,237	1,733	111,773,788	1,708,071	65,438,620
1,294,749	928	95,537	1,928	263,102,581	3,852,529	68,293,466
455,373	633	73,025	641	74,097,974	867,912	85,375,005
206,937	3,226	112,547	2,422	21,399,630	468,578	45,669,319
46,210	8,155	94,991	2,414	23,788,611	461,897	51,501,982
545,094	1,818	102,497	1,782	59,650,532	1,275,252	46,775,485
652,099	7,695	55,337	1,205	27,666,082	324,644	85,219,657
135,555	485	74,367	2,134	82,104,705	1,360,247	60,360,141
290,764		14,410	460	30,705,570	110,920	276,825,308
86,982	775	46,160	1,100	68,329,706	812,577	84,090,144
418,292	4,798	24,717	1,940	9,420,060	72,196	130,479,115
155,172	38	76,372	1,993	19,475,420	515,234	37,799,204
163,023	546	44,856	1,323	16,040,903	266,629	60,161,961
20,264	177	52,383	1,231	11,827,261	272,310	43,433,074
N/A	1,459	63,953	5,908	16,425,523	490,375	33,495,824
82,218	218	111,837	1,045	16,482,481	959,499	17,178,212
130,350	2,936	16,699	287	18,185,911	163,476	111,245,131
143,668	969	44,418	491	14,037,868	340,309	41,250,354
1,727	8	156,324	2,831	35,263,872	3,285,956	10,731,693
6,733	831	84,668	1,901	19,895,734	1,030,676	19,303,585
260,700	1,040	47,893	437	21,424,251	652,399	32,839,174
24,234	274	38,657	703	39,932,569	1,047,548	38,120,030
69,460	1,438	8,812	152	8,686,306	52,150	166,562,572
218,413	1,898	10,318	124	20,666,268	163,974	126,033,534

(Continued)

(Continued)

#	Country, Other	Total Cases	New Cases	Total Deaths	New Deaths	Total Recovered	New Recovered
29	Belgium	1,163,726		25,320		1,083,370	
30	Pakistan	1,127,584	+3,772	25,003	+80	1,012,662	+3,107
31	Sweden	1,116,584		14,629		*1,082,660*	+445
32	Romania	1,091,340	+415	34,425	+13	1,052,055	+68
33	Thailand	1,066,786	+17,491	9,562	+242	861,770	+22,134
34	Portugal	1,020,546	+1,126	17,645	+6	957,359	+1,043
35	Israel	999,110	+5,899	6,856	+26	923,616	+5,798
36	Morocco	813,945	+2,996	11,889	+97	730,669	+6,124
37	Hungary	811,121	+340	30,052	+6	771,530	+1,753
38	Jordan	790,450	+976	10,308	+15	767,785	+914
39	Switzerland	758,984	+6,218	10,944	+3	701,746	+1,649
40	Nepal	748,981	+1,548	10,533	+24	700,097	+1,924
41	Serbia	744,150	+1,837	7,218	+4	716,304	+479
42	Kazakhstan	743,220	+6,314	8,415	+113	615,816	+5,097
43	UAE	710,438	+1,060	2,024	+4	692,585	+1,659
44	Austria	677,603	+1,077	10,763	+1	654,355	+791
45	Tunisia	642,788	+1,891	22,609	+72	592,891	+850
46	Lebanon	592,780	+624	8,014	+3	546,444	+1,581
47	Cuba	592,619	+9,320	4,618	+74	538,043	+7,832
48	Greece	561,812	+2,626	13,385	+34	511,520	+3,439
49	Saudi Arabia	542,354	+360	8,490	+9	529,377	+741
50	Georgia	517,098	+2,354	6,831	+60	453,488	+5,200
51	Ecuador	498,728	+114	32,092	+7	443,880	
52	Bolivia	486,643	+249	18,302	+6	427,631	+770
53	Belarus	470,635	+918	3,691	+10	464,300	+701
54	Paraguay	457,838	+113	15,593	+22	435,386	+207
55	Panama	452,986	+388	7,015	+6	436,866	+581
56	Costa Rica	445,442	+1,480	5,361	+19	357,239	+1,087
57	Bulgaria	443,186	+1,891	18,532	+57	402,268	+626
58	Guatemala	440,007	+754	11,516	+16	379,835	+2,281
59	Kuwait	408,434	+189	2,407	+3	401,837	+520
60	Azerbaijan	394,451	+2,945	5,340	+32	346,810	+1,027
61	Sri Lanka	394,355	+4,355	7,560	+194	344,381	+20,991

Active Cases	Serious, Critical	Tot Cases/ 1M pop	Deaths/1M pop	Total Tests	Tests/1M pop	Population
55,036	178	99,914	2,174	18,240,201	1,566,051	11,647,262
89,919	5,390	4,994	111	17,276,450	76,519	225,779,880
19,295	30	109,779	1,438	11,536,406	1,134,222	10,171,206
4,860	193	57,166	1,803	11,205,580	586,962	19,090,813
195,454	5,615	15,240	137	8,129,670	116,138	70,000,185
45,542	151	100,421	1,736	16,510,990	1,624,666	10,162,698
68,638	664	107,132	735	18,484,044	1,981,991	9,326,000
71,387	1,411	21,758	318	8,660,806	231,511	37,409,854
9,539	10	84,209	3,120	6,492,441	674,029	9,632,291
12,357	632	76,603	999	9,031,711	875,270	10,318,774
46,294	166	86,970	1,254	9,550,767	1,094,404	8,726,914
38,351		25,189	354	3,839,395	129,124	29,734,155
20,628	39	85,563	830	4,915,426	565,180	8,697,105
118,989	221	39,049	442	11,575,012	608,154	19,033,023
15,829		70,854	202	72,200,005	7,200,721	10,026,774
12,485	77	74,748	1,187	74,116,688	8,176,038	9,065,110
27,288	630	53,746	1,890	2,479,635	207,331	11,959,791
38,322	54	87,296	1,180	4,759,054	700,843	6,790,472
49,958	494	52,357	408	7,538,108	665,986	11,318,710
36,907	319	54,204	1,291	14,777,535	1,425,743	10,364,797
4,487	1,127	15,307	240	26,960,332	760,900	35,432,180
56,779		129,911	1,716	7,715,357	1,938,333	3,980,409
22,756	759	27,786	1,788	1,748,358	97,409	17,948,605
40,710	220	41,051	1,544	2,224,611	187,657	11,854,679
2,644		49,825	391	7,674,134	812,441	9,445,769
6,859	171	63,303	2,156	1,755,066	242,663	7,232,519
9,105	108	103,134	1,597	3,592,540	817,932	4,392,224
82,842	427	86,540	1,042	1,968,955	382,527	5,147,233
22,386	223	64,336	2,690	4,044,359	587,111	6,888,573
48,656	5	24,053	630	2,096,054	114,581	18,293,284
4,190	132	94,053	554	3,687,295	849,097	4,342,608
42,301		38,507	521	4,329,920	422,698	10,243,530
42,414		18,329	351	4,575,585	212,661	21,515,852

(Continued)

(Continued)

#	Country, Other	Total Cases	New Cases	Total Deaths	New Deaths	Total Recovered	New Recovered
62	Slovakia	394,093	+11	12,547		*380,512*	+19
63	Uruguay	384,181	+87	6,016		376,848	+137
64	Myanmar	375,871	+2,186	14,499	+125	302,447	+3,572
65	Croatia	369,838	+73	8,303	+2	359,099	+330
66	Vietnam	358,456	+10,280	8,666	+389	154,612	+6,945
67	Dominican Republic	348,026	+191	3,994	+5	339,309	+312
68	Ireland	338,707	+1,590	5,074		*286,796*	+1,345
69	Denmark	338,240	+774	2,567	+1	322,525	+969
70	Palestine	327,634	+1,324	3,642	+5	314,770	+318
71	Honduras	326,830		8,594		105,153	
72	Venezuela	325,716		3,895		311,746	
73	Oman	301,450	+151	4,038	+7	290,410	+266
74	Libya	296,879	+1,625	4,076	+25	214,470	+1,270
75	Ethiopia	296,731	+927	4,571	+10	270,171	+1,039
76	Lithuania	294,086	+314	4,494	+6	276,188	+153
77	Egypt	286,541	+189	16,676	+5	236,075	+211
78	Bahrain	271,715	+84	1,387	+1	269,362	+102
79	Moldova	264,746	+314	6,369	+6	255,485	+206
80	Slovenia	263,664	+116	4,441	+1	255,725	+45
81	Armenia	237,885	+251	4,762	+10	224,493	+239
82	S. Korea	237,782	+1,416	2,222	+7	207,601	+1,325
83	Qatar	231,126	+289	601		227,656	+211
84	Kenya	229,628	+619	4,528	+31	213,473	+1,437
85	Bosnia and Herzegovina	209,909	+127	9,740	+9	192,218	
86	Zambia	204,651	+102	3,578	+4	198,781	+222
87	Mongolia	195,250	+1,566	902	+4	184,559	+1,674
88	Algeria	192,089	+506	5,034	+30	130,351	+401
89	Nigeria	187,588	+565	2,276	+8	168,818	+363
90	Kyrgyzstan	174,148	+228	2,489	+5	166,702	+394
91	North Macedonia	169,202	+330	5,695	+27	152,467	+369
92	Afghanistan	152,660	+77	7,083	+7	108,863	+601

Active Cases	Serious, Critical	Tot Cases/ 1M pop	Deaths/1M pop	Total Tests	Tests/1M pop	Population
1,034	11	72,143	2,297	3,277,867	600,049	5,462,666
1,317	14	110,160	1,725	3,277,102	939,674	3,487,489
58,925		6,856	264	3,435,339	62,658	54,826,464
2,436	42	90,729	2,037	2,470,457	606,055	4,076,294
195,178		3,645	88	15,271,562	155,296	98,338,634
4,723	136	31,719	364	1,932,424	176,122	10,972,102
46,837	60	67,731	1,015	6,415,912	1,282,992	5,000,742
13,148	18	58,162	441	79,807,831	13,723,263	5,815,514
9,222	31	62,568	696	2,074,147	396,099	5,236,438
213,083	552	32,410	852	945,963	93,806	10,084,245
10,075	681	11,492	137	3,359,014	118,510	28,343,772
7,002	81	57,371	768	1,550,000	294,989	5,254,431
78,333		42,546	584	1,475,718	211,485	6,977,892
21,989	510	2,510	39	3,170,408	26,821	118,204,859
13,404	92	109,784	1,678	4,321,090	1,613,083	2,678,778
33,790	90	2,741	160	3,068,679	29,356	104,534,065
966	4	153,620	784	5,796,788	3,277,330	1,768,753
2,892	26	65,804	1,583	1,532,742	380,972	4,023,239
3,498	10	126,806	2,136	1,427,627	686,602	2,079,265
8,630		80,107	1,604	1,489,110	501,456	2,969,575
27,959	399	4,633	43	12,694,029	247,352	51,319,633
2,869	19	82,316	214	2,458,570	875,620	2,807,805
11,627	149	4,166	82	2,316,674	42,029	55,120,231
7,951		64,439	2,990	1,113,991	341,981	3,257,464
2,292	165	10,788	189	2,225,969	117,344	18,969,572
9,789	192	58,485	270	3,647,780	1,092,644	3,338,490
56,704	38	4,292	112	230,861	5,158	44,753,611
16,494	11	885	11	2,648,684	12,496	211,954,964
4,957	131	26,200	374	1,631,560	245,461	6,646,926
11,040		81,219	2,734	1,111,451	533,511	2,083,276
36,714	1,124	3,824	177	748,754	18,753	39,926,713

(Continued)

(Continued)

#	Country, Other	Total Cases	New Cases	Total Deaths	New Deaths	Total Recovered	New Recovered
93	Botswana	150,842	+4,381	2,171	+90	140,350	+6,090
94	Uzbekistan	149,876	+818	1,028	+7	142,632	+887
95	Norway	149,732	+906	811		88,952	
96	Mozambique	143,127	+343	1,808	+8	126,313	+272
97	Latvia	141,118	+73	2,569		136,903	+115
98	Albania	139,721	+397	2,478		131,451	+178
99	Estonia	139,126	+319	1,285	+4	131,596	+43
100	Namibia	123,861	+280	3,345	+3	118,114	+29
101	Zimbabwe	123,001	+349	4,293	+44	107,759	+949
102	Finland	122,417	+371	1,008		46,000	
103	Uganda	118,777	+104	2,960	+8	95,375	+50
104	Ghana	115,102	+518	974	+6	107,391	+422
105	Cyprus	111,666	+333	484	+1	90,755	
106	Montenegro	110,377	+415	1,680	+9	102,271	+206
107	China	94,652	+21	4,636		88,321	+74
108	El Salvador	92,686		2,846	+6	79,029	
109	Cambodia	89,641	+410	1,808	+16	85,618	+537
110	Rwanda	83,519	+496	1,027	+6	45,233	+11
111	Cameroon	82,454		1,338		80,433	
112	Maldives	80,067	+108	225	+2	78,189	+113
113	Luxembourg	75,130	+145	830		73,641	+174
114	Senegal	72,015	+88	1,680	+9	57,889	+495
115	Singapore	66,576	+98	50	+1	65,700	+99
116	Jamaica	62,712	+879	1,402	+14	47,787	+66
117	Malawi	59,624	+153	2,082	+8	45,721	+169
118	DRC	54,009		1,053		30,858	
119	Ivory Coast	53,730	+85	397	+2	52,597	+192
120	Réunion	46,754		310		42,770	
121	Angola	46,076	+131	1,163	+10	42,624	+240
122	Australia	44,920	+892	984	+3	31,593	
123	Fiji	44,188	+591	444	+6	24,425	+616
124	Trinidad and Tobago	43,145	+112	1,237	+12	36,717	+249

Active Cases	Serious, Critical	Tot Cases/ 1M pop	Deaths/1M pop	Total Tests	Tests/1M pop	Population
8,321	1	62,701	902	1,616,148	671,792	2,405,725
6,216	23	4,405	30	1,377,915	40,500	34,022,495
59,969	7	27,375	148	6,950,992	1,270,828	5,469,655
15,006	32	4,438	56	829,467	25,719	32,251,450
1,646	10	75,775	1,379	3,311,552	1,778,188	1,862,318
5,792	3	48,612	862	1,007,108	350,396	2,874,203
6,245	13	104,799	968	1,716,723	1,293,146	1,327,555
2,402	39	47,761	1,290	646,571	249,318	2,593,355
10,949	12	8,141	284	1,231,543	81,507	15,109,719
75,409	23	22,055	182	6,329,278	1,140,295	5,550,560
20,442	527	2,507	62	1,553,673	32,790	47,382,825
6,737	53	3,618	31	1,544,519	48,552	31,811,819
20,427	92	91,725	398	9,366,973	7,694,257	1,217,398
6,426	11	175,716	2,674	590,111	939,432	628,157
1,695	23	66	3	160,000,000	111,163	1,439,323,776
10,811	152	14,208	436	1,192,402	182,783	6,523,601
2,215		5,278	106	2,054,302	120,963	16,982,929
37,259	43	6,271	77	2,322,189	174,369	13,317,665
683	152	3,021	49	1,751,774	64,172	27,297,966
1,653	18	145,208	408	1,302,818	2,362,759	551,397
659	5	117,834	1,302	3,385,705	5,310,160	637,590
12,446	50	4,176	97	720,702	41,790	17,245,985
826	7	11,279	8	16,926,698	2,867,527	5,902,890
13,523	60	21,073	471	533,840	179,386	2,975,937
11,821	285	3,028	106	378,191	19,205	19,692,474
22,098		583	11	306,299	3,306	92,656,867
736		1,981	15	868,523	32,023	27,121,763
3,674	33	51,799	343	173,764	192,513	902,610
2,289	8	1,354	34	830,836	24,416	34,028,370
12,343	90	1,739	38	29,819,169	1,154,102	25,837,560
19,319	310	48,888	491	362,890	401,487	903,865
5,191	22	30,715	881	297,206	211,585	1,404,668

(Continued)

(Continued)

#	Country, Other	Total Cases	New Cases	Total Deaths	New Deaths	Total Recovered	New Recovered
125	Madagascar	42,847	+2	955	+1	41,276	+2
126	Eswatini	40,714	+259	1,022	+18	30,114	+404
127	Guadeloupe	40,018		324		2,250	
128	French Polynesia	39,117	+2,745	311	+54	31,215	+2,039
129	Sudan	37,640		2,826	+4	31,593	
130	Malta	35,831	+57	436		34,312	+37
131	Cabo Verde	34,773	+35	305	+2	33,840	+34
132	French Guiana	33,452	+392	207	+3	9,995	
133	Mauritania	32,026	+204	677	+4	28,420	+253
134	Martinique	31,369		243		104	
135	Guinea	28,802		314		26,212	
136	Suriname	27,574	+71	700	+1	23,308	+50
137	Syria	27,003	+102	1,977	+6	22,309	+22
138	Gabon	25,717	+50	165		25,484	+47
139	Guyana	24,336		594		22,364	
140	Haiti	20,746	+27	583		16,986	+307
141	Togo	19,702	+88	173	+1	16,123	+18
142	Mayotte	19,682	+18	175		2,964	
143	Seychelles	19,594	+204	104	+3	18,909	+259
144	Papua New Guinea	17,838	+6	192		17,547	
145	Bahamas	17,386	+108	338	+8	13,712	+8
146	Somalia	16,892	+105	927	+9	8,141	+64
147	Tajikistan	16,514	+91	124		16,297	+123
148	Taiwan	15,932	+6	828		14,868	+21
149	Belize	15,556	+141	353	+2	14,262	+113
150	Andorra	15,002	+14	130	+1	14,770	+41
151	Curaçao	14,974	+37	138	+1	14,298	+58
152	Mali	14,763	+6	536		14,059	+6
153	Timor-Leste	14,403	+187	49	+1	10,971	+67
154	Lesotho	14,382	+11	400		6,763	
155	Aruba	14,060	+57	128	+2	13,200	+70

Active Cases	Serious, Critical	Tot Cases/ 1M pop	Deaths/1M pop	Total Tests	Tests/1M pop	Population
616	16	1,503	34	237,216	8,323	28,501,130
9,578	10	34,683	871	303,755	258,762	1,173,877
37,444	23	99,995	810	335,167	837,495	400,202
7,591	38	138,338	1,100	26,355	93,205	282,763
3,221		836	63	238,579	5,300	45,014,935
1,083	3	80,903	984	1,164,947	2,630,348	442,887
628	23	61,777	542	205,858	365,723	562,880
23,250	26	108,790	673	350,852	1,141,012	307,492
2,929	16	6,688	141	428,072	89,392	4,788,723
31,022	1	83,666	648	292,898	781,203	374,932
2,276	24	2,127	23	529,762	39,131	13,538,050
3,566	13	46,528	1,181	99,642	168,134	592,634
2,717		1,501	110	103,566	5,759	17,984,602
68	5	11,252	72	1,061,767	464,561	2,285,526
1,378	16	30,770	751	254,453	321,730	790,889
3,177		1,794	50	105,799	9,151	11,561,156
3,406		2,318	20	447,687	52,671	8,499,636
16,543	1	70,216	624	176,919	631,159	280,308
581		197,833	1,050	21,504	217,118	99,043
99	7	1,952	21	150,870	16,506	9,140,169
3,336	16	43,735	850	130,339	327,869	397,534
7,824		1,030	57	192,556	11,743	16,397,581
93		1,688	13			9,781,597
236		668	35	4,672,849	195,795	23,866,055
941	6	38,329	870	210,055	517,560	405,856
102	7	193,809	1,679	193,595	2,501,034	77,406
538	4	90,829	837	224,323	1,360,688	164,860
168		706	26	372,453	17,809	20,914,167
3,383		10,692	36	169,501	125,830	1,347,062
7,219		6,653	185	146,348	67,702	2,161,644
732	14	131,054	1,193	177,885	1,658,076	107,284

(Continued)

(Continued)

#	Country, Other	Total Cases	New Cases	Total Deaths	New Deaths	Total Recovered	New Recovered
156	Burkina Faso	13,715	+2	171		13,461	
157	Congo	13,493	+95	179		12,421	
158	Laos	12,621	+152	11		4,604	
159	Hong Kong	12,063	+5	212		11,766	
160	Benin	12,021	+1,838	125	+6	8,572	+170
161	Djibouti	11,698	+2	157		11,531	+4
162	South Sudan	11,310		120		10,948	
163	CAR	11,270	+19	99		6,859	
164	Burundi	10,791		38		773	
165	Nicaragua	10,672		198		4,225	
166	Channel Islands	10,327	+123	94	+1	9,730	+112
167	Iceland	10,177	+62	30		9,191	+125
168	Gambia	9,470	+31	301	+5	9,049	+184
169	Equatorial Guinea	9,049		123		8,803	
170	Mauritius	8,098	+1,028	25	+3	1,854	
171	Yemen	7,539	+30	1,420	+2	4,670	+26
172	Saint Lucia	7,232	+121	97		5,700	+6
173	Eritrea	6,624	+1	37		6,571	+6
174	Isle of Man	6,366	+54	37		5,877	+112
175	Sierra Leone	6,355		121		4,342	
176	Niger	5,770		196		5,490	
177	Guinea-Bissau	5,518		103		4,592	
178	Liberia	5,459		148		2,715	
179	Gibraltar	5,290	+3	96		5,088	+14
180	San Marino	5,261	+1	90		5,119	
181	Chad	4,987	+2	174		4,805	
182	Barbados	4,652	+12	48		4,454	
183	Comoros	4,051	+1	147		3,893	+5
184	Sint Maarten	3,394		42		3,037	
185	Liechtenstein	3,228	+5	59		3,104	+28

Active Cases	Serious, Critical	Tot Cases/ 1M pop	Deaths/1M pop	Total Tests	Tests/1M pop	Population
83		636	8	209,088	9,700	21,555,316
893		2,379	32	188,207	33,178	5,672,688
8,006		1,706	1	360,617	48,758	7,395,986
85		1,594	28	22,744,169	3,005,926	7,566,443
3,324	5	963	10	604,310	48,405	12,484,520
10		11,647	156	203,299	202,411	1,004,388
242		997	11	209,687	18,485	11,343,503
4,312	2	2,288	20	60,228	12,228	4,925,496
9,980		878	3	345,742	28,124	12,293,443
6,249		1,589	29			6,714,502
503		58,779	535	796,374	4,532,759	175,693
956	21	29,604	87	945,877	2,751,458	343,773
120	3	3,797	121	101,817	40,827	2,493,891
123		6,218	85	187,519	128,847	1,455,358
6,219		6,355	20	358,675	281,495	1,274,181
1,449	23	247	46	221,795	7,254	30,574,093
1,435	4	39,180	526	65,853	356,761	184,586
16		1,839	10	23,693	6,577	3,602,416
452	1	74,415	433	83,753	979,029	85,547
1,892		779	15	160,729	19,693	8,161,798
84		229	8	139,342	5,529	25,201,567
823	4	2,730	51	89,236	44,155	2,020,957
2,596	2	1,051	28	128,246	24,696	5,193,023
106	5	157,071	2,850	330,405	9,810,416	33,679
52	1	154,676	2,646	74,465	2,189,310	34,013
8		294	10	137,185	8,088	16,961,901
150		16,165	167	248,865	864,784	287,777
11		4,548	165			890,701
315	2	78,145	967	37,841	871,270	43,432
65	3	84,386	1,542	49,126	1,284,239	38,253

(Continued)

(Continued)

#	Country, Other	Total Cases	New Cases	Total Deaths	New Deaths	Total Recovered	New Recovered
186	Monaco	3,141	+8	33		3,015	+15
187	Saint Martin	3,094		41		1,399	
188	New Zealand	3,054	+38	26		2,874	
189	Bermuda	2,750		33		2,593	
190	Turks and Caicos	2,613	+1	20		2,503	
191	Bhutan	2,585		3		2,553	
192	British Virgin Islands	2,568		37		2,500	
193	Sao Tome and Principe	2,524		37		2,403	
194	St. Vincent Grenadines	2,320		12		2,261	
195	Brunei	1,873	+104	3		436	+17
196	Caribbean Netherlands	1,776	+2	17		1,721	
197	Antigua and Barbuda	1,540	+50	43		1,313	+14
198	St. Barth	1,489		2		462	
199	Tanzania	1,367		50		183	
200	Dominica	1,339		1		794	
201	Faeroe Islands	997	+2	2		991	+3
202	Saint Kitts and Nevis	823		3		604	
203	*Diamond Princess*	712		13		699	
204	Cayman Islands	663		2		649	
205	Wallis and Futuna	445		7		438	
206	Greenland	305	+7			218	+6
207	Grenada	222		1		182	
208	Anguilla	166				135	
209	New Caledonia	135				58	

Active Cases	Serious, Critical	Tot Cases/ 1M pop	Deaths/1M pop	Total Tests	Tests/1M pop	Population
93	6	79,398	834	54,960	1,389,282	39,560
1,654	7	78,492	1,040	52,859	1,340,986	39,418
154	5	611	5	2,727,853	545,342	5,002,100
124	1	44,344	532	436,268	7,034,879	62,015
90	1	66,463	509	94,789	2,411,014	39,315
29		3,309	4	798,573	1,022,100	781,306
31	7	84,305	1,215	77,252	2,536,095	30,461
84		11,278	165	14,689	65,635	223,797
47	2	20,837	108	64,768	581,698	111,343
1,434	27	4,235	7	204,923	463,352	442,262
38		67,016	641	20,707	781,367	26,501
184	9	15,578	435	17,409	176,106	98,855
1,025		150,237	202	38,369	3,871,355	9,911
1,134	7	22	0.8			61,669,290
544		18,548	14	41,298	572,074	72,190
4		20,316	41	401,000	8,171,167	49,075
216	1	15,346	56	28,471	530,888	53,629
0						
12		9,955	30	124,033	1,862,273	66,603
0		40,407	636	20,508	1,862,163	11,013
87	2	5,362		55,170	969,886	56,883
39		1,963	9	54,402	480,927	113,119
31		10,953		37,111	2,448,601	15,156
77		468		41,120	142,465	288,632

(Continued)

(Continued)

#	Country, Other	Total Cases	New Cases	Total Deaths	New Deaths	Total Recovered	New Recovered
210	Falkland Islands	66				63	
211	Macao	63				59	
212	Saint Pierre Miquelon	30				30	
213	Vatican City	27				27	
214	Montserrat	25		1		20	
215	Solomon Islands	20				20	
216	Western Sahara	10		1		8	
217	*MS Zaandam*	9		2		7	
218	Vanuatu	4		1		3	
219	Marshall Islands	4				4	
220	Samoa	3				3	
221	Saint Helena	2				2	
222	Micronesia	1				1	
	Total:	213,273,985	+511,053	4,453,036	+7,564	190,809,781	+570,821

Source: Worldometers.info

Active Cases	Serious, Critical	Tot Cases/ 1M pop	Deaths/1M pop	Total Tests	Tests/1M pop	Population
3		18,359		7,409	2,060,918	3,595
4		96		4,739	7,186	659,433
0		5,207		7,800	1,353,697	5,762
0		33,624				803
4		5,005	200	1,408	281,882	4,995
0		28		4,500	6,373	706,058
1		16	2			614,056
0						
0		13	3	23,000	72,942	315,317
0		67				59,646
0		15				199,920
0		328				6,098
0		9				116,399
18,011,168	112,251	27,361.1	571.3			

This is the recent vaccination data worldwide as of 7th June 2021.

Doses given	Fully vaccinated		
2.12 billion	458 million		
Location	**Doses given**	**Fully vaccinated**	**% of population fully vaccinated**
United Kingdom	67.3M 67,300,000	27.2M 27,200,000	40.8% 40.8%
China (Mainland)	763M 763,000,000	-	-
United States	300M 300,000,000	138M 138,000,000	42.1% 42.1%
India	226M 226,000,000	44.6M 44,600,000	3.3% 3.3%
Brazil	71.4M 71,400,000	22.9M 22,900,000	10.8% 10.8%
Germany	54.2M 54,200,000	17.2M 17,200,000	20.8% 20.8%
France	39.6M 39,600,000	12.4M 12,400,000	18.5% 18.5%
Italy	37.7M 37,700,000	12.9M 12,900,000	21.4% 21.4%
Mexico	34.5M 34,500,000	14.1M 14,100,000	11.1% 11.1%

Russia	30.6M 30,600,000	13.1M 13,100,000	9.1% 9.1%
Turkey	30.6M 30,600,000	13.1M 13,100,000	15.9% 15.9%
Spain	28.8M 28,800,000	10.3M 10,300,000	21.9% 21.9%
Indonesia	28.7M 28,700,000	11.1M 11,100,000	4.1% 4.1%
Canada	25.8M 25,800,000	2.71M 2,710,000	7.2% 7.2%
Poland	22M 22,000,000	8.17M 8,170,000	21.5% 21.5%
Chile	19.4M 19,400,000	8.39M 8,390,000	44.3% 44.3%
Japan	15.6M 15,600,000	3.96M 3,960,000	3.1% 3.1%
Morocco	15M 15,000,000	5.9M 5,900,000	16.2% 16.2%
Saudi Arabia	14.8M 14,800,000	-	-
Argentina	13.7M 13,700,000	3M 3,000,000	6.7% 6.7%
United Arab Emirates	13.3M 13,300,000	3.84M 3,840,000	39.3%

Source: Our World in Data

BIBLIOGRAPHY

Sharma, S., Bloom, D., Buchan, L. & Milne, O., 2021. *Chronicle Live.* [Online] Available at: https://www.chroniclelive.co.uk/news/uk-news/boris-johnson-coronavirus-crisis-mistakes-19708170 [Accessed 19 05 2021].

Allegretti, A. & Walker, P., 2021. *Boris Johnson: inquiry into Covid response will start in spring 2022.* [Online] Available at: https://www.theguardian.com/world/2021/may/12/boris-johnson-inquiry-into-handling-of-covid-crisis-will-start-spring-2022

Ani, 2021. *Countries relying on Chinese Covid vaccines reporting surge in infections.* [Online] Available at: https://www.business-standard.com/article/current-affairs/countries-relying-on-chinese-covid-vaccines-reporting-surge-in-infections-121062300082_1.html [Accessed 24 06 2021].

Anon., 2020. [Online] Available at: https://apiject.com/

Apiject, 2021. *Solving drug delivery challenges.* [Online] Available at: https://apiject.com/

Ashford, J., 2020. *The fatal viruses that the world has learned to live with.* [Online] Available at: https://www.theweek.co.uk/107001/the-fatal-viruses-that-the-world-has-learned-to-live-with [Accessed 31 05 2021].

Baibhawi, R., 2021. *COVID-19: Europe Battered With 3rd Wave, France Tops Infection List As UK Gears For Unlock.*

[Online] Available at: https://www.republicworld.com/world-news/uk-news/covid-19-europe-battered-with-3rd-wave-france-tops-infection-list-as-uk-gears-for-unlock.html [Accessed 01 07 2021].

Barczyk, H., 2020. *Eight Persistent COVID-19 Myths and Why People Believe Them.* [Online] Available at: https://www.scientificamerican.com/article/eight-persistent-covid-19-myths-and-why-people-believe-them/ [Accessed 27 05 2021].

Brewer, K., n.d. *Parosmia: Since I had Covid, food makes me want to vomit'.* [Online] Available at: https://www.bbc.co.uk/news/stories-55824567

Brewer, K., n.d. *Parosmia: 'Since I had Covid, food makes me want to vomit'.* [Online] Available at: https://www.bbc.co.uk/news/stories-55824567

Burke, J., 2021. *South Africa tightens Covid rules as 'devastating wave' gathers pace.* [Online] Available at: https://www.theguardian.com/world/2021/jun/27/south-africa-expected-tighten-covid-rules-third-wave-gathers-pace [Accessed 30 06 2021].

Business Standard, 2021. *Scientists in China discussed weaponising coronavirus in 2015: Report.* [Online] Available at: https://www.business-standard.com/article/current-affairs/scientists-in-china-discussed-weaponising-coronavirus-in-2015-report-121050900916_1.html [Accessed 06 06 2021].

Carl Emmerson, T. W. A. X. X., 2020. *A response to the Chancellor's new package of support.* [Online] Available at: https://www.ifs.org.uk/publications/14764

Carlsen, A., Huang, P., Levitt, Z. & Wood, D., 2021. *How Is The COVID-19 Vaccination Campaign Going In Your State?.* [Online] Available at: https://www.npr.org/sections/

health-shots/2021/01/28/960901166/how-is-the-covid-19-vaccination-campaign-going-in-your-state

Cher, A., 2020. *US, Italy and Spain have the most coronavirus cases. These charts show their infection curves.* [Online] Available at: https://www.cnbc.com/2020/04/01/charts-show-the-coronavirus-spike-in-us-italy-and-spain.html

Clinic, C., 2021. *Cleveland Clinic Post.* [Online] Available at: https://health.clevelandclinic.org/lose-sense-of-smell-covid-19-anosmia/

Fifth Sense, 2020. [Online] Available at: https://www.fifthsense.org.uk/

France-Presse, A., 2021. *WHO could have sounded COVID-19 alarm 'sooner', says independent global panel report.* [Online] Available at: https://www.firstpost.com/health/who-could-have-sounded-covid-19-alarm-sooner-says-independent-global-panel-report-9614921.html/amp [Accessed 24 05 2021].

Grech, S. & Cuschien, S., 2020. *COVID-19: a one-way ticket to a global childhood obesity crisis?.* [Online] Available at: https://www.ncbi.nlm.nih.gov/pmc/articles/PMC7644278/

Hannah Ritchie, *Coronavirus (COVID-19) Vaccinations.* [Online] Available at: https://ourworldindata.org/covid-vaccinations?country=OWID_WRL [Accessed 07 06 2021].

Helmore, Edward; Guardian Newspaper, 2021. *US authorizes Pfizer coronavirus vaccine for children ages 12 to 15.* [Online] Available at: https://www.theguardian.com/world/2021/may/10/pfizer-vaccine-fda-authorized-children-12-15

Heren, K., 2020. *Spain's daily coronavirus death toll rises again despite recent decline in Covid-19-linked fatalities,* s.l.: Evening Standard.

History.com, 2020. *Pandemics That Changed History.* [Online] Available at: https://www.history.com/topics/middle-ages/pandemics-timeline

International Monetary Fund, 2020. *IMF World Economic Outlook, April 2020: The Great Lockdown.* [Online] Available at: https://www.imf.org/en/Publications/WEO/Issues/2020/04/14/weo-april-2020

Joe Marshall, A. N. C. H. R. H., 2020. *Coronavirus Act 2020.* [Online] Available at: https://www.instituteforgovernment.org.uk/explainers/coronavirus-act

JULIANA MENASCE HOROWITZ, A. B. A. R. M., 2021. *A Year Into the Pandemic, Long-Term Financial Impact Weighs Heavily on Many Americans.* [Online] Available at: https://www.pewresearch.org/social-trends/2021/03/05/a-year-into-the-pandemic-long-term-financial-impact-weighs-heavily-on-many-americans/ [Accessed 29 05 2021].

Karaban, E., 2020. *World Bank Fast-Tracks $128 Million COVID-19 (Coronavirus) Support for Sri Lanka.* [Online] Available at: https://www.worldbank.org/en/news/press-release/2020/04/01/world-bank-fast-track-support-covid19-corona

Keep, M., Brien, P. & Harari, D., 2021. *Coronavirus: Economic impact.* [Online] Available at: https://commonslibrary.parliament.uk/research-briefings/cbp-8866/

Khalil, S., 2021. *Australia Covid: Seventh city locks down amid vaccine chaos.* [Online] Available at: https://www.bbc.co.uk/news/world-australia-57661144 [Accessed 03 07 2021].

Khan, H., 2020. The mental health emergency. *How has the coronavirus pandemic impacted our mental health,* pp. 9-15.

Maas, S., 2020. *Social and Economic Impacts of the 1918 Influenza Epidemic.* [Online] Available at: https://www.nber.org/digest/may20/social-and-economic-impacts-1918-influenza-epidemic

Ma, C. M., Cong, Y. P. & Zhang, H. M. P., 2020. *COVID-19 and the Digestive System.* [Online] Available at: https://journals.lww.com/ajg/fulltext/2020/07000/covid_19_and_the_digestive_system.11.aspx

Marshall, L., Bibby, J. & Abbs, I., 2020. *Emerging evidence on COVID-19's impact on mental health and health inequalities.* [Online] Available at: https://www.health.org.uk/news-and-comment/blogs/emerging-evidence-on-covid-19s-impact-on-mental-health-and-health#:~:text=More%20than%20two%2Dthirds%20of,and%20feeling%20bored%20(49%25).

Maas, S., 2021. *Social and Economic Impacts of the 1918 Influenza Epidemic.* [Online] Available at: https://www.nber.org/digest/may20/social-and-economic-impacts-1918-influenza-epidemic

Mguni, M., 2021. *Botswana Secures Enough Vaccine Doses For Its Adult Population.* [Online] Available at: https://www.bloombergquint.com/onweb/botswana-secures-enough-vaccine-doses-for-its-adult-population

MPH Online, 2021. *Outbreak: 10 of the worst pandemics in history.* [Online] Available at: https://www.mphonline.org/worst-pandemics-in-history/ [Accessed 23 05 2021].

National Geographic, 2019. *Genetically Modified Organisms.* [Online] Available at: https://www.nationalgeographic.org/encyclopedia/genetically-modified-organisms/ [Accessed 30 05 2021].

National World War 2 Museum, 2020. *Medical Innovations: From the 1918 Pandemic to a Flu Vaccine.* [Online] Available at: https://www.nationalww2museum.org/war/articles/medical-innovations-1918-flu

Nirmita Panchal, 2021. *The Implications of COVID-19 for Mental Health and Substance Use.* [Online] Available at: https://www.kff.org/coronavirus-covid-19/issue-brief/the-implications-of-covid-19-for-mental-health-and-substance-use/

Nyenswah, T., 2020. *Africa has a COVID-19 time bomb to defuse.* [Online] Available at: https://www.weforum.org/agenda/2020/04/africa-covid-19-time-bomb-defuse/

Nyenswah, T., 2020. *Africa has a COVID-19 time bomb to defuse.* [Online] Available at: https://www.weforum.org/agenda/2020/04/africa-covid-19-time-bomb-defuse/

Pandey, A., 2021. *Rich countries block India, South Africa's bid to ban COVID vaccine patents.* [Online] Available at: https://www.dw.com/en/rich-countries-block-india-south-africas-bid-to-ban-covid-vaccine-patents/a-56460175

Pandey, A., 2021. *Rich countries block India, South Africa's bid to ban COVID vaccine patents.* [Online] Available at: https://www.dw.com/en/rich-countries-block-india-south-africas-bid-to-ban-covid-vaccine-patents/a-56460175

Reynolds, E., 2020. *How it all went wrong (again) in Europe as second wave grips continent.* [Online] Available at: https://edition.cnn.com/2020/09/19/europe/europe-second-wave-coronavirus-intl/index.html

Ryan, D. D., n.d. *Coronavirus (COVID-19) & Obesity.* [Online] Available at: https://www.worldobesity.org/news/statement-coronavirus-covid-19-obesity [Accessed 04 06 2021].

Shaban, A. R. A., 2020. [Online] Available at: https://www.africanews.com/2020/07/29/coronavirus-in-africa-breakdown-of-infected-virus-free-countries//

Shearing, H., 2021. *Covid: PM to push for world vaccination by end of 2022.* [Online] Available at: https://www.bbc.co.uk/news/uk-57373120 [Accessed 06 06 2021].

Sky News, 2021. *COVID-19: Brazil's coronavirus crisis worsens as patients 'tied to beds' and ventilated without sedatives.* [Online] Available at: https://news.sky.com/story/covid-19-patients-tied-to-beds-and-intubated-without-sedatives-as-coronavirus-crisis-worsens-in-brazil-12276739

Slater, S., 2021. *Universal Credit statistics, 29 April 2013 to 14 January 2021.* [Online] Available at: https://www.gov.uk/government/statistics/universal-credit-statistics-29-april-2013-to-14-january-2021/universal-credit-statistics-29-april-2013-to-14-january-2021 [Accessed 29 05 2021].

Usan Lund, 2021. *The future of work after COVID-19.* [Online] Available at: https://www.mckinsey.com/featured-insights/future-of-work/the-future-of-work-after-covid-19

Vedyashkin, S., 2021. *Coronavirus in Russia: The Latest News June 28.* [Online] Available at: https://www.themoscowtimes.com/2021/06/28/coronavirus-in-russia-the-latest-news-june-28-a69117 [Accessed 29 06 2021].

Wimbledon, 2021. *The Championships 2021 - Latest Updates.* [Online] Available at: https://www.wimbledon.com/en_GB/about_wimbledon/the_championships_2021_latest_update.html [Accessed 25 06 2021].

World Bank, 2020. *The Global Economic Outlook During the COVID-19 Pandemic: A Changed World.* [Online] Available at: https://www.worldbank.org/en/news/feature/2020/06/08/

the-global-economic-outlook-during-the-covid-19-pandemic-a-changed-world

World Health Organisation, 2020. *Obesity and overweight.* [Online] Available at: https://www.who.int/news-room/fact-sheets/detail/obesity-and-overweight#:~:text=For%20adults%2C%20WHO%20defines%20overweight,than%20or%20equal%20to%2030. [Accessed 06 06 2021].

Worldometer, 2021. *Reported Cases and Deaths by Country or Territory.* [Online] Available at: https://www.worldometers.info/coronavirus/ [Accessed 02 06 2021].

Lightning Source UK Ltd.
Milton Keynes UK
UKHW010636280122
397869UK00001B/93